ISBN: 9781407790695

Published by:
HardPress Publishing
8345 NW 66TH ST #2561
MIAMI FL 33166-2626

Email: info@hardpress.net
Web: http://www.hardpress.net

MACHINE MOLDER
PRACTICE

AN INSTRUCTIVE, ILLUSTRATED MANUAL ON
MOLDER WORK—THE OPERATION AND
SUPERINTENDANCE OF THE
MOLDING MACHINE

BY

W. H. ROHR
**PROFESSIONAL WRITER ON MACHINE
WOOD-WORKING**

$2.50

INDIANAPOLIS
PRACTICAL BOOKS CO.
INDIANA

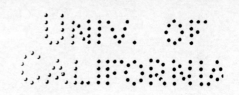
PREFACE.

Several years previous to the publishing of this book, the writer, then a practicing woodworker, was firmly convinced that a manual on molder practice would be welcomed by thousands of machine woodworkers who possess only the limited knowledge acquired by years of experience in a single or several establishments.

Molder work involves many operations and considerable technique. A book which would lack any of the details or variety of modern methods used in all kinds of woodworking plants would be incomplete and of limited value to the trade in general.

Undoubtedly the previous absence of such a work as this is accountable for in the fact that to produce a comprehensive treatise most likely to meet the requirements of the greatest number of persons meant months of traveling for the author, and consultation with hundreds of practical men over the country in order to uncover and separate the most modern and efficient methods in use.

Fortunately, the writer has had just such an opportunity to gather material for this book, hence the knowledge disclosed in the subsequent pages has been verified by personal observation and practical experience.

No attempt is made to establish in each case set rules for the subject treated. Unusual conditions require special treatment, and in numerous occasions one may, even with the aid of this book, be compelled to lean upon his own skill in solving the problems of his work. Some of the principles and practical details, while superfluous to the expert, are included, however, for the general class.

The author desires to record in these pages special acknowledgement of the assistance rendered him by Mr. G. H. Oburn, and to machinery manufacturers who permitted the illustrating of their molders and equipment. Thanks, too, for the courtesies extended to me by friends thruout the United States whose suggestions and practical ideas assisted in making possible this, the first work on molder practice.

W. H. ROHR.

CONTENTS.

A typical four-side, square-head molder.

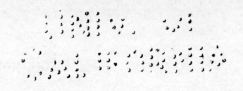

CHAPTER I.

THE ALIGNMENT OF A MOLDER.

The four-sided molder has held an important place in the wood-working industry ever since wood-working machines came into general use. Today the modern improved molder is a greatly used machine in most factories manufacturing wood products; in fact, it is absolutely indispensable in establishments making interior and exterior wood trim, fixtures, show cases, cabinets, furniture, pianos, picture frames, caskets, incubators, harvesting machines and other agricultural implements, street and railway cars, toys and novelties, etc.

Altho the bulk of the work done on an ordinary molding machine consists of moldings or molded work of different kinds, the machine's usefulness is not limited to this class of work. A molder can be set up to perform such operations as plain surfacing, gang ripping, plain milling, glue jointing and some kinds of irregular shaping.

It is a most interesting machine. To operate a molder or to even stand near one and observe how rough stock enters the rolls, passes thru the machine and comes out so smooth and nicely molded, or milled to shape, is indeed, fascinating. It is this fascination that has lured many ambitious and mechanically-inclined young men to choose molder work as their vocation. The molder appeals to the average young man, more so, perhaps, than any other wood-working machine because the work gives opportunity to display his mechanical ability, and it possesses enough variety to make it both agreeable and intensely interesting.

There are many things to know about molder work, and heretofore there have been so few occasions to learn even a fair part of them thoroly that today there is a scarcity of first-class moldermen—men who can put a molder in good

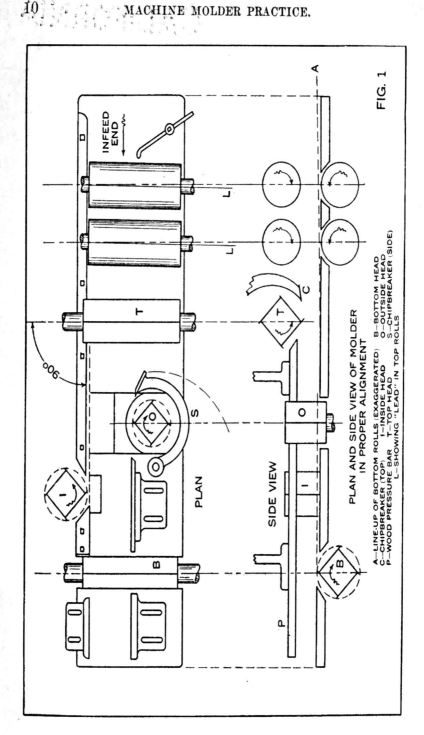

INFEED END

L

L

T

C

T

S

I

90°

B

P

PLAN

SIDE VIEW

A

FIG. 1

PLAN AND SIDE VIEW OF MOLDER
IN PROPER ALIGNMENT

A—LINE-UP OF BOTTOM ROLLS (EXAGGERATED)   B—BOTTOM HEAD
C—CHIPBREAKER (TOP)   I—INSIDE HEAD   O—OUTSIDE HEAD
P—WOOD PRESSURE BAR   T—TOP HEAD   S—CHIPBREAKER (SIDE)
L—SHOWING "LEAD" IN TOP ROLLS

Typical inside molder. It will be noticed that the top head and feed rolls of so-called inside molders have the two-side drive. Inside machines are often used as a combined molder, planer and matcher.

order, keep it in perfect repair, make and temper cutters, and set up accurately and quickly for any kind of molding or milled work.

The care and operation of a molder should be learned thoroly, step by step, beginning first with the alignment of the machine. Correct machine alignment is a necessary requirement in the production of good molding. No amount of skill at setting up or feeding the machine will successfully overcome imperfections in the line-up of the bed, guides, chipbreakers, feed rolls, etc. It is the imperfect alignment of some parts of the machine that causes a great deal of apparently mysterious molder trouble. Testing and adjusting the line-up of a molder, when necessary, is a comparatively simple matter if once clearly understood. A long and short straight edge, a steel square, and wrenches to make adjustments are needed before starting.

Begin by testing and leveling up the bed. See that the bed plates opposite the side heads are neither too high nor too low, and that the rear table back of the bottom head lines up with the main bed lengthwise, and that crosswise it lines with the cutting circle of the bottom head. Level the lower feed rolls with the bed and let their upper faces come above the bed just enough to relieve the friction of stock as it passes over the bed under the top infeed rolls. The first bottom roll should be set slightly higher than the second to permit the stock to feed straight into the machine without cramping or bending. The exact amount of elevation recommended for the lower rolls, see dotted line A in Fig. 1, will depend upon the kind of stock generally run, and whether it is surfaced or run in the rough. Ordinarily the second roll is given 1/32 to 1/16-in. elevation above the bed and the first roll is set slightly higher.

As to the guides, those parts which form the inside guide or guide rail from the infeed end of the machine to the inside head I, are the only really permanent, stationary guides on a molder (all others are adjustable as occasion requires, to the finished width of the molding) therefore, they must be lined with special care. When they are adjusted to a perfectly straight line from end to end, and set square with the top head, the bolts should be set down tight so that no part of the guide rail can possibly shift or move out of position.

By studying the line-up of the feed rolls in the plan of Fig. 1, it is apparent that they are not set perfectly square with the guide rail. A slight forward lead is given them so that the stock feeding thru the machine will have a tendency to always feed tightly up to the guide rail. This is an important detail. It assists the machine feeder wonderfully, especially when he has crooked lumber or short lengths of wide material to run. All molding machines do not permit of this adjustment but on the later types of molders the adjustment is generally possible. When the rolls are set in this manner it is unnecessary to put excessive tension upon the side springs, or to employ extra levers and devices to hold the stock up tightly to the guide rail.

The top head chipbreaker C, and the side head chipbreaker S, should be adjusted to swing in a little past line with the cutting circles described by the top and outside heads respectively, so that both will hold the stock firmly in place even tho it be a little under size in places.

It is also apparent in. Fig. 1, that the inside ends of the top and bottom heads are set a fraction of an inch "in" past the line of the guide rail, and similarly the lower ends of both side heads fall a little below the bed line. This is done purposely to permit of bolting overhanging molding cutters close to the ends of the heads, and when the heads are so positioned to suit the operator, they should be marked and not shifted from their position unless absolutely necessary.

The side head spindles should, of course, set plumb with the machine bed, and the top and bottom heads should line parallel with it. If they do not, the defect should be corrected or the amount they are "out" should be carefully measured so that it can be taken into account when making set-ups or adjusting the molder gage.

With these suggestions, any molder can be lined up correctly, and, in the language of shop, the molderman "knows where he is." Operation can then begin with the assurance that the machine is in condition to turn out first-class molding insofar as alignment is concerned.

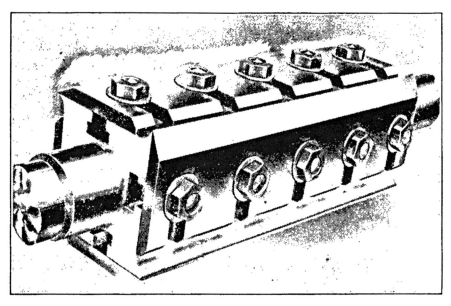

Typical square, slotted cutterhead fitted with ordinary straight surfacing knives made of carbon steel.

CHAPTER II.

PLANNING SET-UPS AND SELECTING KNIVES.

When ready to set up a molder for any kind of work, first examine the drawing or sample furnished, then plan the set-up accordingly. The golden rule to observe is to choose the quickest and easiest method by which smooth, accurate molding can be produced with safety. The best way to handle any particular job is always governed by such local conditions as the size and kind of machine and cutters available, amount of molding to be run, the kind and condition of stock, whether soft or hardwood, wet, green, semi-dry or kiln dried, surfaced or unsurfaced, etc.

All stock for moldings should be thoroly kiln dried before it is worked and all hardwoods for high-grade finish moldings should be surfaced before being put thru the molder. Softwood may be run in the rough, but if high-grade finish is required, it is best to have the stock planed first, unless it is to be manufactured into molding on a modern five-head or six-head molder having straight planer heads at the front and molding profile heads at the rear end.

An expert, when acquainted with his machine and the conditions which prevail where he works, can tell the instant he sees the outlines of any ordinary molding just how he is going to make it. A less experienced man, however, must give a little time to studying it out.

Moldings are now run both face up and face down, but the old-established practice is to make them face up, therefore the face-up method will be discussed now, and the face-down system explained in Chapter VII. All of the early four-side molders were designed to work the face side of molding with the top head, the edges with the side heads, and the back with the bottom head, and this is the way moldings are worked at the present time in a large number of factories.

Another well-established practice which has been handed down by pioneers in the trade is to work the thick edge of moldings next to the guide rail, and this is usually the best plan even in light of modern methods, unless one has to run two or more different widths having exactly the same molded side and edge, as for example, round-edge casings, chair rail, apron, base, etc. In the latter case the molded edge should be run next to the guide rail regardless of

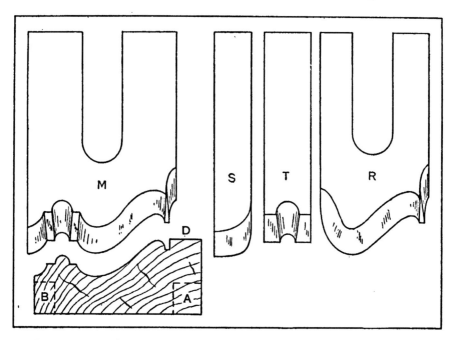

Fig. 2. M, wide, solid knife. S, T, R, sectional knives for combination set-up.

whether it is thick or thin, so the knives will not have to be shifted or changed when the machine is changed to suit different widths of the same style of molding. With this information it is apparent that moldings similar to the one shown in Fig. 2 should be run with the molded side up and the thick edge to the guide rail. The edge will be surfaced with the side heads and if the stock has not been previously planed, the back must be dressed with the bottom head.

If a rabbett is required as indicated by dotted lines at A, it can be cut with knives either on the inside head or bottom head. A bevel, cove or quarter-round can be made at this

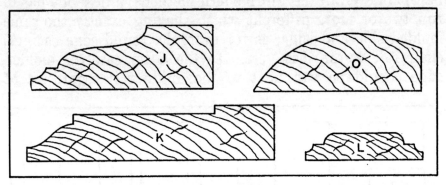

Fig. 3. Patterns with thin edges are often worked with the top head.

corner in the same manner, but in most cases it is best to use the bottom head for such work because it is more accessable, has greater belt power, and there is more clearance for the sweep of cutters. Tongues and grooves and certain other shapes, however, can only be worked on the edges with the side heads. When a rabbett is made in such manner that a very thin edge is left on the molding, as at B, Fig. 2, the rabbett should be worked with the side head instead of the bottom head. Otherwise, the thin remaining edge would probably be broken off, or at least, badly chipped and split away in places.

When a pattern tapers down to a comparatively thin edge as at J, K, L and O, Fig. 3, it is generally better to leave the outside head idle, or both side heads if both edges are thin, and finish the edge as well as the face of the molding with the top head, provided the stock is not more than 1⅛-in. thick. It is usually dangerous to cut down thru stock thicker than 1⅛-in. with ordinary cutters.

Double beveled edges like those on crown moldings, see Fig. 4, are made with knives on the top and bottom heads whenever possible—rarely with the side heads. However, when making patterns like this the outside head (carrying straight knives) is used to size the stock to a uniform width so that it will advance properly between close-fitting guides to the bottom head, where the final cut is made. Comparatively thin moldings, 13/16-in. or less, are often sized to width with a pair of sizing knives on the top head.

A point worth remembering in planning work ahead at the molder is the possibility of saving time and labor by grouping the work according to its kind and size. Hardwood moldings and those containing deep, heavy cuts are run at slow feeds and often with special cutters, therefore these should be grouped whenever possible. Likewise there can be a grouping of widths and thicknesses and patterns of the same profile. This always helps to materially reduce the amount of work necessary in changing a machine from one pattern to another.

SELECTING KNIVES.

The types of molding knives commonly used on square heads consist of slotted knives, see M and R, Fig 2; spike knives, S and T, Fig. 2; milled-to-pattern slotted knives, X, Fig. 5; thin high-speed steel knives under caps, Y and Z, Fig. 5. Wide slotted knives, made of either solid or laid carbon steel, are the old-fashioned type and are still used in many plants. At one time it was the universal custom to make a pair of wide knives for each pattern of molding, but this slow, expensive practice has been largely superseded by what modern moldermen call the combination set-up in which several narrow knives are combined to make any required cut. The latter is by far the better method to use when a large variety of moldings are manufactured, and it is often the best on stock work. Some of the chief reasons why the wide solid knives have been abandoned are, because they are expensive in first cost, difficult to make, and when once made they are hard to keep in shape and sharpen. Each wide knife is only good for one pattern of molding. Take the knife M, Fig. 2, for example; it cannot be used to make any other pattern, hence

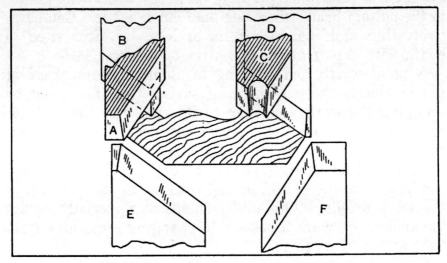

Fig. 4. Showing how sectional cutters should be arranged.

its usefulness is very limited. It contains two corners which can be reached only with a three-cornered file, thus necessitating a filing temper which will not hold a sharp edge very long. In the process of resharpening, this knife will gradually wear wider between the corners that are filed, and in doing so it loses its original shape. If a small nick developes at any point, the entire edge must be reground and filed in order to preserve the same general profile.

Narrow knives, on the other hand, are less expensive in first cost, easier to make, and can be so designed that all inside corners are eliminated, thus permitting the entire edge to be sharpened on a grinding wheel. This feature allows using a harder temper, which holds a keen edge longer, and makes it possible to keep the knives in correct shape indefinitely. Narrow knives also have the important advantage of being readily adapted to a wide range of work on a variety of patterns. Another point in their favor is that by breaking up a wide, complicated cut with several narrow knives (see S, T and R, Fig. 2, and A, B, C, D, etc., Fig. 4), distributed around on different sides of the head, there is less strain on the bolts and a better balanced and an easier-running cutting unit is provided. The result is a smoother finish and less chipping of the grain.

A large assortment of narrow knives is extremely valuable in plants manufacturing odd and special work because the same knives can be used repeatedly in different combinations to cut all sizes and shapes of molding. New patterns can be worked and obsolete ones frequently matched without making or changing a cutter. The necessary combinations are simply built up with the knives at hand. Narrow knives should be grouped in the racks according to shape and size. An assortment for general work will include a large variety of quarter and half-rounds, coves, beads, ogees, reverse ogees, bevels, surfacers, grooving and rabbetting cutters, quirks, etc.

Spike knives like S and T, Fig 2, are made from long bars of carbon or self-hardening steel, 1/4 to 5/16-in. thick. Often they are spread at the end to make a wider cut. They are fastened to the head under square steel caps or washers about 1/4-in. thick, see Fig. 7, Chapter IV. Two spikes must be placed under a plain cap like that in Fig. 7, but caps can be made with one edge bent over or offset in such a manner that the shoulder will take the place of the slug shown in Fig. 7. Caps of the latter type are especially good for side heads where there is seldom any call for more than one molding knife on each side of the head.

Milled knives like X, Fig. 5, may be purchased from stock or made in the grinding room on a modern head-grinding machine. The profile of the mold is milled in the back, and a face bevel is ground on the front as shown.

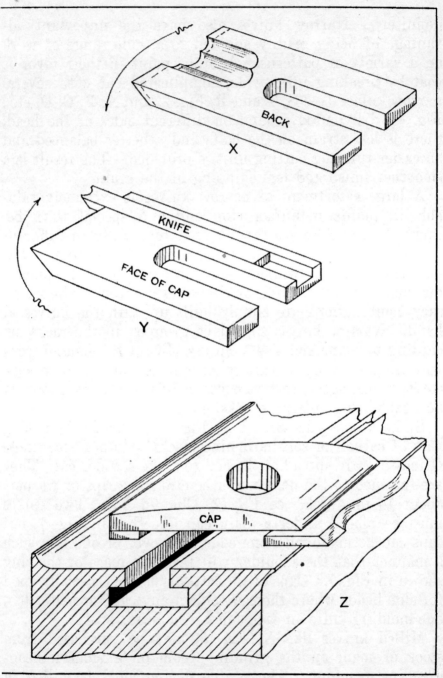

Fig. 5. X is milled knife with front bevel. Y and Z show thin steel knives under caps.

The knife is sharpened by grinding the flat face-bevel. In some cases a strip of high-speed steel is brazed onto the front beveled face, as shown by dotted lines at X, Fig. 5, in order to give a harder and longer-lasting cutting edge. The advantages of milled knives are that they retain their correct shape until worn out, are easily sharpened, produce very smooth work on kiln-dried stock, and never tear out cross grain. On the other hand, their usefulness is limited to comparatively light cuts and considerable more power is required to drive them thru the stock. Knives of this type are used in a great many mills in the Pacific Northwest for working bone-dry fir.

The cutter shown at Y, in Fig. 5 consists of an ordinary slotted knife, milled as shown, to receive a narrow strip of high-speed steel. The chief purpose of this kind of combination is to utilize thin-steel knives that are worn too narrow for further use as surfacing knives. The narrow cutters are particularly adapted to making beads and V's on beaded and V-ceiling, also other light, narrow cuts of a similar nature.

Still another method of using thin high-speed steel on square slotted heads is shown at Z. High-speed steel cutters will hold an edge for a long time and are quite satisfactory for relatively light cuts. They are often made from broken planer, matcher or jointer knives, altho in some plants, regular high-speed bar steel is purchased for making small molding and spike knives. Unlike carbon steel, high-speed steel requires no heat treatment before using.

CHAPTER III.

BALANCING MOLDER KNIVES.

Correct knife balance is one of the most important factors in molder work. Every molderman who expects to produce high-grade molding and keep his machine in good running order, with the least amount of trouble and expense, should exercise particular care and good judgment in balancing the cutters.

It is a well-known mechanical axiom that any object revolving at high speed must be in a perfect state of balance to proceed smoothly and safely without vibration. This is especially true of the cutterheads on molding machines, as they run from 3,000 to 4,000 r.p.m., and any slight deviation from a perfect running balance produces a violent jar, causing the bearings to heat, hence a wavy finish on the surface of the molding. There are at least four kinds of balance that must be observed when balancing a set of knives: dead-weight, line, projection and thickness balance. Theoretically, the knives on opposite sides of a cutterhead should pair exactly in dead-weight, in thickness and in projection, and they should be in perfect line directly opposite one another on the head. It is quite possible to keep straight dressing knives and some molding knives in this theoretically ideal alignment, but when using sectional knives it is more often impossible and even undesirable to always have them paired in this manner.

In actual practice, especially in factories turning out odd detail work where frequently twenty or more set-ups are made in a day on a machine, there is seldom, if ever, more than one knife used for each member of the molding. As a result, a knife that cuts one member is balanced by one that cuts another member, or by a "dead" knife bolted opposite to it. Right here it might be well to explain that on short runs there is no advantage in bolting a pair of knives shaped exactly alike onto opposite sides of the heads with the expectation that both knives will cut alike and consequently produce smoother work than one, because it is

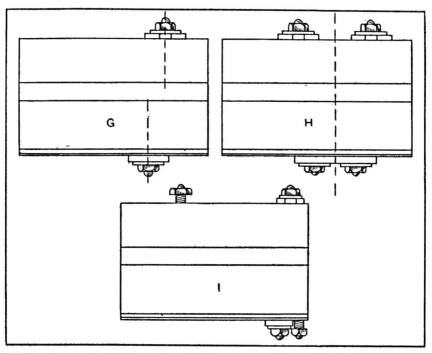

Fig. 6. G, knives set in staggered fashion. H and I, correcting staggered effect.

practically impossible to set two knives to cut alike. They can be made to cut alike by jointing at full speed or by repeatedly whetting the knife that sets out the farthest, but this is only profitable on long runs.

To successfully balance sectional knives one must strive to become an expert at balancing, which he can do by using good judgment and observing the tried-and-tested rules of knife balance. Perfection in balancing is only acquired by experience, but the following rules will serve as a valuable guide:

Rule 1. When necessary to mate two knives, one of which must be set out farther than the other, the knife projecting the farthest should be the lightest in dead weight. Any deficiency in weight in the short knife can be made up by using an extra washer or heavier bolt, or by slipping a slug of metal in the bolt slot under the short knife. It is always a good plan to keep a number of washers and weight slugs of different sizes on hand for use in 'forcing a balance in emergencies. The same methods are followed to produce a balance when one knife is thicker or heavier than its mate.

Rule 2. The center of weight of a pair of knives, bolted to opposite sides of a cutterhead, should come exactly in line. In other words, no pair of knives should be set in staggered fashion as shown at G, Fig. 6. In practice this rule cannot always be followed to the letter, so the next best thing under adverse conditions is to observe Rule 3.

Rule 3. The center of weight of a group of knives, bolted to opposite sides of the head, must come absolutely in line to maintain the running balance of the cutterhead. This latter rule is a hard and fast one—there is no getting around it. However, it permits setting a pair of knives in staggered fashion on condition that suitable weights or an extra knife or set of knives, as required, are so placed as to counteract the staggered effect and bring the center of weight on opposite sides of the head in line, see H, Fig. 6. One more method to counteract the effect of staggered knives appears at I, Fig. 6.

Rule 4. Still another point in knife balance that should be observed when using sectional cutters is to distribute the cutting knives around the head as evenly as possible. In other words, do not bolt all the knives on one side of the head and the balance weights on the other side. Likewise, do not bolt all the knives that cut heaviest on one side, and those that cut lightest on the other side if it can be avoided. Furthermore, when running a many-membered molding, avoid putting both knives, which cut at or near the outer edges of the molding, on the same side of the head. Let one edge-knife come on one side of the head while the other comes on the opposite side of the head. These last points are clearly illustrated in Fig. 4, Chapter II. Here the top bevel knives A and D are not placed on the same side of the head, neither are the bottom bevels E and F. Knives A and C on one side of the head are balanced by B and D, respectively, on the opposite side, while the knife that cuts the middle part is balanced by a mate of the same shape or by any knife of the right weight that comes handy. Bevels E and F are balanced with similar bevels of the same general shape and weight, or by any available knives of the required weight.

Thin, high-speed steel surfacing knives attached to square head with caps and bolts.

CHAPTER IV.

SETTING UP A MOLDER.

Setting the knives and adjusting the machine for a run of new molding are tasks that test a molderman's skill and ability. It is therefore to his interest to have things handy about the machine and to practice a method of setting up that is at once simple, accurate and rapid. The common mistake of some operators is to proceed with, the execution of their work with no well-defined or systematic method, and as a result, they make a number of false moves and do more or less unnecessary tinkering at the machine every time they set it up. These useless moves consume valuable time and reduce efficiency. Even by the best methods there are several adjustments to make when setting up for new work or changing from one job to another, therefore it is apparent that a predetermined and efficient system is very essential.

The first requisite for convenience in setting up is to have a variety of wood pressure bars grouped together in a rack near the machine. The necessary machine wrenches and a long-handled screw driver should also have their appointed places within easy reach. Extra knife bolts with washers and nuts, screws for the pressure bars, hammer, pliers, bevel and try-square, oil-slip, wiping waste and a few clean blank templets should be kept in a drawer near at hand.

The machine should be fitted with. index plates and pointers to show at a glance just how far to set the top head from the bed of the machine to cut a given thickness and how far to set the outside head from the guide rail to cut a given width. There should also be marks to indicate the location of the bottom and inside heads for different size cuts. It requires an extra move and additional time to measure machine adjustments with a rule and the chances for inaccuracy are much greater. It is not safe to determine these distances by counting the turns made with the crank or hand wheel because machine screw threads develop play in time, and when a screw is reversed it sometimes requires half a turn or more before the head will move.

If the feed is not under perfect control and cannot be stopped instantaneously with the feed lever, a simple quick-acting brake of home-made construction should be added for this purpose. It is also helpful, if quick machine stops are desired, to add a brake to the machine countershaft because the natural momentum of any smooth-running molder will often cause the cutterheads to continue in motion for some time after the power has been turned off. The blower pipes connected to the hoods should be arranged to telescope or swing out of the way in such a manner that they will not have to be replaced every time the operator changes or sets the knives on a head.

To proceed in setting up be sure to have all knives, etc., selected, balanced, and laid out carefully before you begin to put them on the heads. Then make it a point to set each head complete as you go; for example, start with the top head, and follow in regular order to the outside head, inside head and finally the bottom head. By adopting and adhering to a systematic routine of this sort the work soon becomes easy and natural, and, one's movements get to be automatic, so to speak, resulting in remarkable speed and accuracy.

There are a number of methods used and exploited for positioning or setting knives on square cutterheads, but the majority of uptodate moldermen thruout the country use the molder rule in some form for this purpose. Molder rules made of metal or celluloid can be bought for a nominal price, but if one prefers he can make a rule, being careful, however, to leave off all patented features of rules now on the market. The ordinary molding rule is lined off in ⅛-in. divisions both ways, see Fig. 7, but the eighths by which the projection or overhang of the knives is measured are longer than true ⅛-in. divisions.   The reason for this is that a molder knife bolted to a square

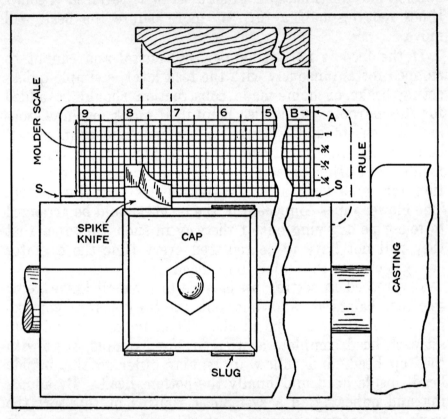

Fig. 7. Method of using molder or "stickerman's" rule. A is in line with guide rail. B is in line with inside head.
S is surfacing line.

cutterhead strikes the work and does its cutting at an angle which varies according to the size of the head and amount of knife projection.

Referring to Fig. 7, line S, on the rule, is the surfacing line to which the cutting edge of the straight surfacing knives is set.   The edge of the rule rests against the lip of the head, and surfacing line S is scribed parallel to, and about 3/32-in., or ⅛-in. from this edge. A knife edge or point set up to the next line above the surfacing line will cut exactly ⅛-in. deep, and if set to the second line above and parallel to line S, it will cut ¼-in. deep, etc. Lines running lengthwise of the rule and parallel to line S are spaced to represent the molder scale, while those crossing them are spaced exactly ⅛-in. apart. Line A indicates the position of the guide rail, and the distance from line A to B shows the amount allowed for the cut of the inside head, therefore the width and location of cuts to be made by knives on the top head are measured from line B unless the inside edge of the mòlding requires no surfacing, in which event the cuts must be measured from line A. The method of using the rule in actual practice is shown clearly in Fig. 7. The working edge of the rule always rests against the lip of the head while the end is either butted against a convenient journal-box casting as shown or arranged to hook over the end of the cutterhead as indicated by dotted line. The former method is recommended when the construction of the machine permits its use, because when the rule rests against an immovable casting, the head can be shifted laterally either way by the adjusting wheel without moving the rule, consequently any lateral shifting does not destroy the alignment of line A with the guide rail nor line B with the inside head. On the other hand, if the rule is hooked over the end of the head the relative position of the head must always remain the same. If shifted, the rule will also move and then require readjustment every time to keep it in correct alignment with the guide rail.

One of these molder rules can be arranged to serve for all four heads on a machine if so desired, but it is much easier and more satisfactory to use a shorter rule for the side heads. The principle of using the rule on any of the other heads is the same as described for the top head. Some moldermen, instead of using a rule similar to that in

Fig. 7, use a blank rule having only the lines S, A and B. On this blank they locate the important points of any required knife profile by using an ordinary rule and a separate molder scale. The molder scale in such cases is sometimes lined off on a miniature brass T-square which is just long enough to span the width of the blank rule. Sometimes the scale is marked off on one edge of a regular two-foot, four-fold rule. Others lay off knife projections by measuring with a common rule and making proper allowance for the sweep of the knives. The latter method, however, is rather a hit-and-miss one, serving the purpose fairly well but not· recommended.

Another way to set molder knives, and one that is used successfully by some very fast set-up men, is to make a tracing of the molding upon transparent paper and then fold and fasten this tracing in the proper position on a blank molder rule, as shown in Fig. 8. Line L K is drawn thru the deepest cut and parallel to the face of the molding. The depth of this cut is measured and then laid ,off on the rule according to the molding scale as at L-1 and K-1. The tracing is then clamped to the rule so that line L-K comes exactly over line L-1, K-1, and the thick edge of the molding comes exactly in line with the inside head cut as shown. Now the knife edge that cuts the deepest member will come right over the line K-1, L-1, as shown, while the edge that cuts the flat surface comes only to the surfacing line S on the blank rule. The remaining profile of the molding knife falls below the tracing in proportion to the difference between the actual measurement and the molder scale, as shown in Fig. 8. •

A beginner on the molder would probably not have much success with this method, but an expert who has a good eye for shapes and can tell at a glance the exact allowance to make between the profile of a molding and the corresponding profile of the knife edge to cut the molding, will find that this gives just the guide needed for setting knives quickly and accurately. The method to use, of course, is largely a matter of choice and the success of any particular system will depend altogether upon the man who uses it. The method just described may not appeal to some because it is not generally known to the trade. However, molder-men in some detail interior trim factories are making re-markable speed with it, and prefer this system to any other.

As mentioned at the beginning of this chapter, the methods of knife setting described in the foregoing apply particularly to new moldings which the operator has never run. For stock moldings and repetition work it is not necessary to go thru the process of locating the required position of the knives on a molder rule after the first set-up has been made, provided that set-up has been properly "carded," or traced off on blank templets and filed away for future use. Templets are undoubtedly the best and quick-est for setting up a machine to make moldings which are run repeatedly. The kind of templet recommended is simply a blank rule made of light-colored wood or celluloid and, like the molder rule, is arranged to either butt against a journal box casting as in Fig. 7, or hook over the end of the head as in Fig. 8. After an original set-up is made correctly with a molder rule, the outline and position of the knives as they set on the head are then carefully traced on the templet with a sharp pencil. The size, name and num-ber of the molding are marked on the templet or set of templets, if more than one are necessary for a set-up, and these are filed away in regular order in a rack or drawer. It is a good plan to shellac wood templets after they are marked as the shellac preserves the sharp, clear-cut lines and keeps the templets clean.

When all knives are positioned on the heads, the chip-breakers at the top and outside heads must be adjusted to swing-in close to the point where the knives leave the work. There must be a safe clearance, of course, so the tip of the chipbreaker will not strike the points of the knives or swing into the cutting circle when any over-size stock enters the machine. Before a piece is started into the machine the outer guides and springs are pulled out to clear it and generally the blower hoods are left off for the trial start. The selection of the first piece of material to run

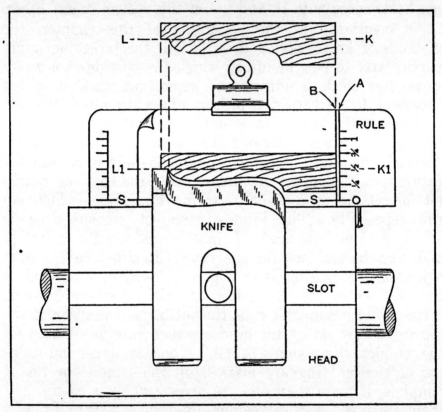

Fig. 8. Another method of setting knives correctly.

is very important. It should be full size, flat, and perfectly straight on one edge. The straight edge is placed next to the guide rail. This piece is fed in slowly and carefully to the top head and past it a few inches, after which the machine is stopped so that a wood pressure bar can be fitted to the machine pressure shoes and as closely as possible to the cutting circle of the top head. Some moldermen fit the pressure bar to the pressure shoes before starting their machine but this is somewhat risky because the end of the bar must be high enough to clear the advancing molding and at the same time be close up to the top head knives. The least miscalculation or even the vibration of the machine may cause the knives to catch the suspended bar and draw it into the head or hurl it back at the operator, thus causing a serious accident. This danger is not so great, of course, when running flat work. The pressure bar should be at least one-half to two-thirds the width of the molding and its underside should be smooth and shaped to conform to the general contour of the molding so that it will perform its function of holding the stock down firmly to the machine bed without marring the finished surface in any way.

With the pressure bar fitted into place and screwed down lightly, the molding is fed past the side heads, after which the side guides and rear end of the pressure bar are adjusted. The molding is then fed over the bottom head, after which the tail board and the remaining guides are adjusted. For the reason that considerable adjusting is necessary while the first piece advances thru the machine, it is best to use a piece of cheap scrap wood for the trial set-up in order to avoid spoiling any good material. Even after all the knives are set correctly the molding will often fail to come exactly true to pattern until the pressure bar and all guides and the tail board back of the bottom head are adjusted and secured firmly in position.

In all cases it is important to have the throat space or gap at the cutter heads as small as possible so the chips will be broken off close and so the stock can neither spring up, down, or sideways. The pressure bar and guides must not be set too tight, however, or the stock will not feed freely. When the molded end of the trial piece is sticking out at the rear of the molder it should be compared to the sample or drawing to make sure it is correct in size and shape. If the machine has been put in proper alignment, the molderman has kept in mind the relative position of the heads with the machine bed and guides, and the molder rule has been adjusted correctly and the set-up made faithfully. as recommended in this chapter, the molding will be right at the first trial. Otherwise, more or less adjusting may be necessary to make it right. Here, let it be emphasized, is where time is lost and troubles begin if any preliminary work has been slighted.

After the set-up is proved correct, the blower hoods are then put in place, the feed regulated if necessary and the machine oiled. The operator can then proceed to feed material into the machine. In feeding, be careful to observe the direction of the grain and look out for defects in the stock. The best side must be used to make the face side of molding and the grain should always be favored if possible, so the knives cutting the face of the molding will not be working against the grain. Each piece of stock that is fed into the machine should be butted squarely against the piece preceding it and the stock should be kept moving forward while the machine is in motion. The rate of feed must, of course, be regulated according to the size of the cut and the kind of stock being run. Generally speaking, hardwoods require a slow or medium feed while softwoods may be run at a faster rate. In taking heavy cuts the feed should be slower than when taking light cuts. One must rely largely upon his own judgment in this matter. When a large hard knot or burly place is approaching the knives, slow down the feed with the lever until the hard place has passed the cutterheads. The yokes, carrying cutterheads, must be securely locked in place so the heads cannot shift or quiver while the machine is in operation. The tension on the top feed rolls should only be sufficient to carry the stock freely thru the machine, because excess tension is hard on the feed mechanism. Experience, naturally acquaints the operator with numerous other little precautions to take while operating a molder.

CHAPTER V.

MAKING UNDER-CUTS AND DOVETAIL GROOVES.

The cutting of moldings which have members that can only be reached by under-cutting presents a different and sometimes more difficult problem than is ever encountered in making ordinary straight-molded cuts. Under-cutting would be almost impossible without special attachments if it were not for the adjustment on modern molders which permits tilting the side heads to any angle up to about 45 degrees.

By working a pattern face up and cutting all of the top profile that can possibly be reached with knives on the top head, an under-cut member can usually be reached with a long, slender knife bolted to the outside head when the latter is tilted to the proper angle. For example, in Fig. 9, the combined cornice and picture molding is worked face up, the top side being almost finished, excepting the under-cut, before it reaches the side head, see A, Fig. 9. This leaves only a very light under-cut to be finished with a knife K on the outside head which must be tilted over to the angle shown. When a molding is worked in this manner the wood pressure bar must be partly cut away just opposite the outside head in order to give room for the swing of the long, overhanging side-head cutter K. The moldings D and E in Fig. 10, which form the sliding frames of one kind of adjustable window screen, are worked in the same manner as just described, the small under-cuts, C C, being made with a knife bolted to the outside head, which is tilted as required.

The small under-cut part of window sills, Fig. 11, for special water-tight frames can be cut in this manner also, provided the sills are not too wide. A very long, slender knife is required to reach point C on the sill. If an under-cut of this kind cannot be reached with an overhanging knife on a tilted sidehead on account of the construction of the machine, the usual alternative is to make all cuts ex-

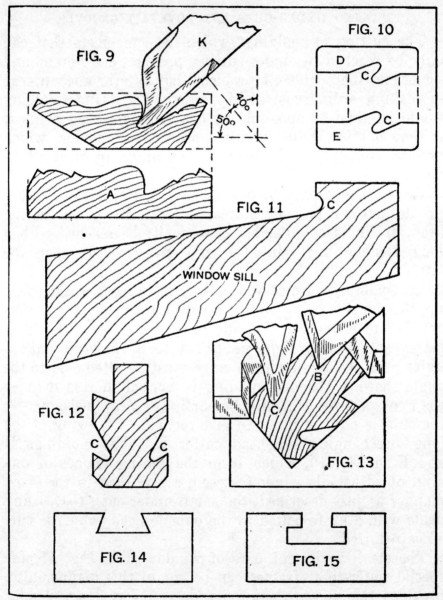

Figs. 9, 10, 11, 12, 13, 14 and 15 show different kinds of moldings with under-cuts.

cepting C on the molder and then do the under-cutting on a shaper.

In running the greenhouse sash pattern, Fig. 12, in one operation thru a molder, it is necessary to tilt both side heads to make the inclined channels, C C. In case the inside head will not tilt enough to make this cut, and, if there is only a comparatively small amount to run, the material can be put thru the machine twice in order to cut both side channels with the tilted outside head. If the latter method is followed, the top and outside heads should be the only ones in action during the first run. The top head should carry surfacing knives to lightly dress the top surface, thus making the stock uniform in thickness so that a pressure bar can be used to hold it firmly in place while passing the outside head. Knives on the outside head should be set to surface the side and mill the inclined channel C. They will, of course, remain unchanged during both runs. On the last run the top part of the molding must be finished with the top head and the bottom part with the bottom head.

In an emergency, this particular pattern, Fig. 12, can be run without tilting or even using the side heads. To do this the stock should first be surfaced on one side and one edge and then run in a trough as shown in Fig. 13. Two runs are required and all cutting must be done with the top head. In place of the regular feed rolls, spur wheels should be used and positioned on the feed shaft to engage the stock just where the deep cuts come, i. e., at points B and C, Fig. 13.

There are many other kinds of under-cuts such as dovetailed "ways" in table slides, dovetailed grooves in Byrkit lath, dovetailed staves, etc., which can be run on a molder. However, work of this kind is generally turned out in large quantities by factories especially equipped for it. Special attachments and sometimes entire machines of exclusive design are built to manufacture such work.

There are table-slide machines which make such patterns as Fig. 14 and Fig. 15, complete in one operation.   One particular machine has, in addition to the four ordinary heads, a pair of overhead tilting arbors mounted near the rear in such manner that they can be fitted with cutters to under-cut the corners of a square-edge groove worked by the top head and make a dovetailed groove of it, as in Fig. 14.   Another machine has overhead vertical spindles, the

Fig. 16.   Attachment for working Byrkit (dovetail) lath, at rear end of an inside molder.

ends of which are fitted with dovetailed router bits to cut pattern Fig. 14, or the ends can be fitted with small milling cutters to make the slot in pattern Fig. 15.

In factories where production is sufficient to justify the expense, special attachments are used which fit on or in line with an ordinary molding machine.   For instance, in making Byrkit lath some mills use a portable stand fitted with tilting spindles which carry either saws or cutterheads, and when the occasion demands it this stand is placed in line with the rear of the machine to make the under-cuts after the straight grooves have been milled with the top or bottom heads. One of these lath-making attachments is shown lined up at the rear on an inside molder in Fig. 16.

An ingenious device for making dovetailed grooves on one edge of a certain pattern consists of a small horizontal spindle attachment mounted just back of the outside head. The end of this spindle carries a dovetail routing bit so positioned that it lines up exactly with a square groove worked in the stock with straight knives on the outside head. Therefore, when the machine is in operation the dovetail router bit finishes the inner corners of the groove cut by the outside head, and makes a perfect dovetail groove. The spindle attachment has a pulley which is belted to the machine countershaft.

CHAPTER VI.

THE USE OF SPECIAL GUIDES AND FORMS.

There are a number of different styles and kinds of molding which, to run thru a molder successfully, require special guides and forms to hold the material in place as it passes thru the machine. One pattern of molding and the forms required for running it has already been shown in Fig. 13, Chapter V. Other patterns requiring the use of forms include wood-split pulley bushings, tapered column staves, piano fall boards, sprung crown molding for circle work, circular church seating, etc. Ordinary small moldings are sometimes run in a simple wood form, when made on large machines, because the form holds the thin narrow strips in place and prevents them from buckling and breaking while passing thru the machine.

The manufacture of wood-split bushings requires absolute accuracy and for that reason it is always advisable to make them in at least two operations instead of one. There is a choice of two different methods however, one being to work the stock into a perfect half-round first and mill the channel last, while the other is just the reserve. In doing the work by the former method, the stock should first be faced off perfectly flat on one side and the two corners should be slabbed off on a saw or molder in order to lighten the finish cut. Then set up the top head with heavy divided knives as shown in Fig. 17, being careful that the extreme ends E, E of the knives are comparatively wide in proportion so there will be no possibility of them chattering or breaking in the cut. Fig. 17 shows the knives on only one side of the head, a quarter-round knife being to the right and a sizing knife S to the left. On the opposite side of the head the relative position of the knives is reversed, therefore it is important that the quarter-round knife on one side be made to balance the sizing knife S on the other side and vice versa. It may be necessary to use a narrow false guide of wood next to the guide rail in order to swing such wide-edged knives but this can easily be arranged if conditions demand it. Should the knives have a very long reach it is advisable to reinforce them with knife braces. Chapter IX explains and illustrates the use of knife braces.

The half-rounds when finished are run in a half-round form, see F, Fig. 18, in order to cut the channel for the shafting. The form F is iron or hardwood, lubricated on the inside with parafine or grease, and it extends from the in-feed end of the machine to a point some distance beyond the top head. It must fit the half-round molding accurately and be securely bolted to the bed of the machine. The friction between the half-round material and this form F is

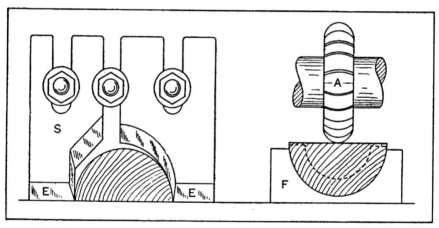

Fig. 17. Knives on one side of head for working heavy, half-round molding. Fig. 18. Form F and spur wheel A, for feeding half-rounds to make pulley bushings.

naturally great, so great in fact that smooth feed rolls will not feed the material along. Sharp corrugated rolls or spur wheels like A, Fig. 18, should be used and there must be plenty of weight or spring pressure employed in feeding. The wood pressure bar back of the top head must be adjusted to work as close as possible to the cutting knives and it must be set down fairly tight to prevent the finished bushings from chattering as they leave the machine. This is a very particular piece of work and accuracy is the thing to keep in mind. The least variation in the thickness of

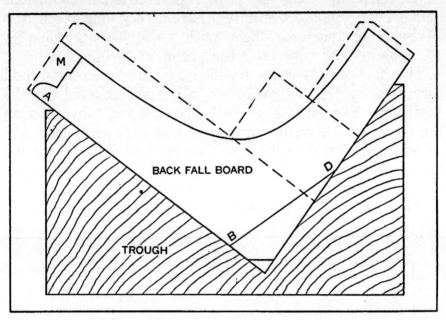

Fig. 19. Trough or form, thru which piano fallboards are fed to top head.

the finished bushings or their shape will render them worthless. When running half-round bushings by the other method, that of cutting the channel first, there is one advantage gained. While the channel is being worked with the top head the outside corners of the stock can be slabbed off with the bottom head, thus saving one operation. On the second run, when the outside surface·is being worked, the stock is run on top of a half-round form or "saddle" which fits the channel perfectly. This form is firmly fastened to the bed of the molder and lined up with the top head knives.

Sometimes pulley bushings are run in quarters. The stock is ripped into squares and then run in a V-trough, the top head cutting the channel and square edges, and the bottom head cutting the convex side.   It is difficult to get an absolutely uniform thickness by cutting one side with the top head and the other side with the bottom head but a slight inaccuracy in quartered bushings does not cause as much harm as in half-round bushings.

Molding the back and front fall boards of pianos is also a particular piece of work and one that requires the use of

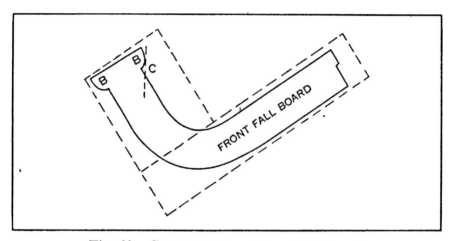

Fig. 20.   Cross-section of front fallboard.

a trough for the first run and a "saddle" for the second run. Fig. 19 shows the cross section of a back fall board in the proper V-trough for the first run thru the machine.

Dotted lines indicate the manner in which the stock is built up and glued before being worked.   The top head is set up with a pair of heavy solid knives to cut the entire concave surface.   As a rule the concave cut is the only one made during the first run because these boards are generally veneered on the concave surface and along the flat surface A B.   Side A B is sometimes veneered before the stock is built up.   The concave surface is veneered immediately after the boards are run thru the molder.   The molded hinge rabbett M is worked on either the molder or shaper. After the fall boards are veneered bevel B D is cut either on a shaper or on the molder. When cut on a molder the fall board is run on a saddle or form having a convex surface to fit the concave side. A pair of bevel knives are bolted on the top head to do the cutting.

Fig. 20 shows a typical front fall board which is generally run in practically the same manner as described for back fall boards. When the board is to be veneered the molded beads, B, B are left off and a molded strip is glued to the surface afterward. In case the front fall board is finished from the solid wood it is completely cut on the molder in two runs with the exception of the under corner C which is worked afterward on a shaper. Corrugated feed rolls or spur wheels like A, Fig. 18, should be used on the feed shaft when work of this kind is run. It is best to use a pair of spur wheels for the second run and position them on the feed shaft so they straddle the hump of the fall board. The saddles or forms over which fall boards and such work are run are made of solid wood, molded accurately to proper shape and bolted securely to the machine bed. The saddle should line up parallel with the guide rail and should extend well past the top head so the pressure bar can be used to good advantage.

Tapered staves for tapered wood-stave columns are often run on ordinary molders, altho there are special molders with moving side heads guided by cams for this purpose, see Fig. 21. There are different methods of running tapered staves but forms are required in every case, unless a special machine or attachment is employed, and the forms must pass thru the machine with the staves. In order to keep the staves moving thru the machine continuously there should be three sets of forms so that while one form is in the machine the helper is taking one out and passing it back, and the feeder is starting another into the machine.

If only plain bevel-edged staves are required they can be completed in one operation by using sets of forms like those shown in Fig. 22. There are no lugs on these forms as the

Fig. 21. Special machine for tapering and jointing column staves at one operation without forms. 1, feed chain; 2, chain dog; 4, chain guideway; 5, sprocket; 6, transmission gear; 7, feed pulley; 8, feed lever; 9, feed roll; 10 and 11, pressure bars; 12, top head; 14, side head; 16, bottom head; 17, mounting shaft; 18, cam; 19, cam traveler; 20 and 21, adjusting screws.

points of a few sharp nails sticking thru them give sufficient grip on each stave to prevent any slippage. When
staves are run in this manner the material should first be
surfaced to uniform thickness, since the top head is idle
during the tapering process. Only the side heads are used
and they are fitted with knives wide enough to cut the full
thickness of a set of forms containing a stave.

In running tapered staves on which a tongue and groove
joint, or any other style of joint, with the exception of the

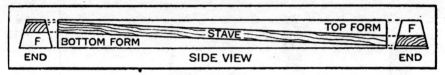

Fig. 22.   One method of making plain tapered staves in one
operation.

plain joint mentioned, is required, the staves are put
thru the molder twice, working only one edge each time.
Fig. 23 shows a simple plan for laying out staves and obtaining correct bevels. One must plan to use lumber thick
enough for the work and always locate the tongue and
groove joint, or whatever style of joint is used, as near
as possible to the inner side of the staves so that when the
columns are turned the turning tool will not cut into the
irregular part of the joint. The number of staves to use
in a column will depend largely upon the size of the
columns and the thickness of the lumber available. The
stock for making staves may be either rough or surfaced
and ripped either straight or tapered, but it must be cut to
an even length. If the stock is surfaced to an absolutely
uniform thickness the top head need not be used. Otherwise it must be used on the first run to bring all staves to
the same thickness. Always run staves with the narrow or
inner side to the top and work the beveled edge of tapered
staves with the outside head. The side head need not be
tilted unless the profile of the joint requires it. If the

stock for staves is not ripped to a taper, the staves may be run the first time without tapered forms. Simply run the tongue edge without taper but to correct bevel. Then with a set of forms corresponding to the thickness, length

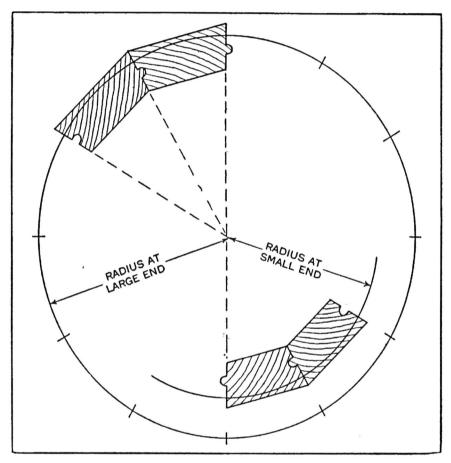

Fig. 23. Showing how staves for tapered columns are laid out.

and taper of finished staves, see Fig. 24, put the stock thru for the final run, this time cutting a groove to fit the tongue. During this final run the top head need not be used but the pressure bar should be set down moderately tight to hold the staves firmly to the bed of the machine as they pass the side head. On the other hand, if the stave stock is ripped to a taper then tapered forms must be used for both runs. The forms should each have a lug at the end, see L, Fig. 24, or sharp nail points along the edge next to the stave in order to prevent any slippage while passing thru the machine. At the beginning of the final run the staves should be tried for correct bevel just as soon as enough are finished to make a column. Accuracy is very important in stave work as the slightest deviation in bevel amounts to considerable when multiplied by the number of staves required to make a complete column.

When straight staves are run on a molder both side heads and the top head are used and one run thru the

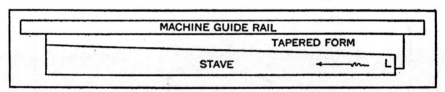

Fig. 24. Stave in position alongside tapered form.

molder completes the job. In this work, when the first good stave is thru the machine there should be enough short sections cut from it to make a complete circle of the required column. The slightest inaccuracy can then be detected and corrected in time to avoid any spoiled work. When the correct bevel and joint are obtained a set of short sections should be assembled into a complete circle and tied securely with a string. This will serve as convincing evidence that the staves fit perfectly and it should be kept by the molderman until after the staves have been glued up.

Sprung crown or cornice molding for circular porches, towers, etc., is another kind of work which at first sight appears difficult, if not impossible to make on an ordinary molder, but in reality it is very simple when proper forms are used. The fact that this kind of molding sets at an angle and is sprung around a circle makes it necessary to treat it as a narrow flat strip bent around a large cone. There-

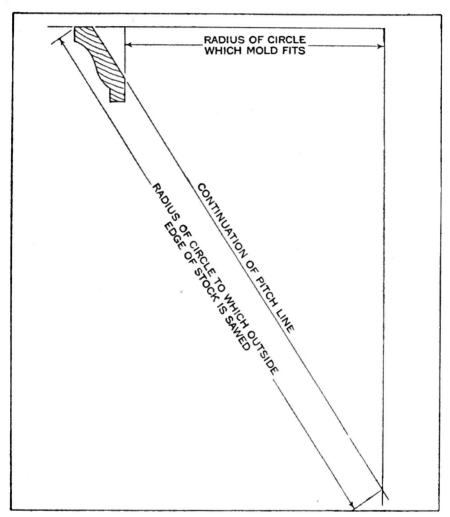

Fig. 25. Method of laying out sprung crown molding for cornice of circular porch or tower.

fore, it must be sawed to a certain radius to make it line up level and fit properly when sprung into place around the circle. The method of finding the correct radius for sawing the stock appears in Fig. 25. The back of the molding (the pitch line) is simply continued until it strikes a line dropped from the center of the circle and the distance from this intersection to the farthest edge of the molding is the radius which must be used in sawing out the stock, see Fig. 25.

In Fig. 26, A, B and C, respectively, are shown three ways to run flat circular work such as that just described. In each case the edge is run against a circular form or guide of wood which is clamped solidly to the machine bed. It will be observed that in both illustrations, A and B, Fig. 26, the inside of the circle is run next to the form, the only difference in the two being that at A the form is next to the guide rail while at B it is on the outside. This difference is immaterial, either method may be used to suit one's convenience since the principle is the same in each. Both methods are recommended and whenever possible either A or B method should be used in preference to that shown at C. The reason why A and B are preferred is because the feed rolls can be used to good advantage in getting the material thru the machine if the circle is not too small in radius. The action of the rolls at A and B tends to crowd the material up tightly to the form as it feeds thru, whereas at C just the reverse is the case. In attempting to feed circle work by method C the rolls carry it straight forward and away from the form. However, method C is sometimes the only alternative, as in the case of very wide material such as circular church seating, etc. It is frequently necessary when running a few wide pieces to the guide as shown at C, to raise the rolls and feed the work by hand.

When running flat circular work of this kind the top head is usually the only one used altho sometimes one of the side heads can be brought into action if necessary. In making crown molding like Fig. 25, the material is put thru the machine twice, the bevels on the back being worked the first time and the face the last time. Circle

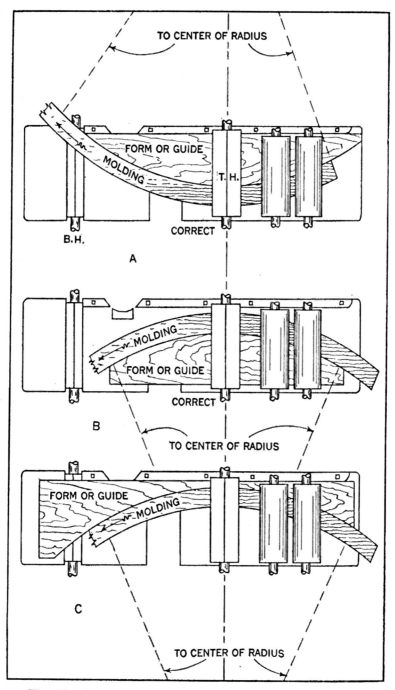

Fig. 26.   Arrangement of guides for running circle work.

work can also be made on edge, that is, the concave or
convex side or edge molded with the top head by the use of
proper forms, see Fig. 27. This class of work is really
shaper work and is seldom run on a molder except in
emergencies.

In a certain large car shop the segment heads for street
car windows are run on a small sash sticker in forms as

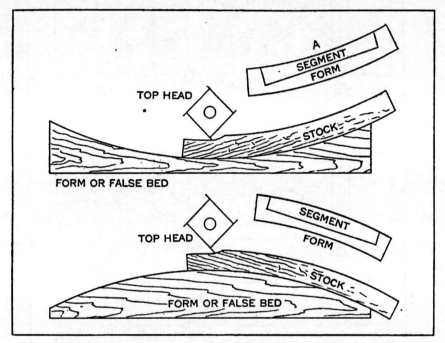

Fig. 27. Molding segment and circle work on edge.

shown at A, Fig. 27. The sticker is used especially for
this work and instead of having regulation square heads it
is fitted with circular, milled, 4-wing cutters which retain
their correct shape until worn out. These cutters do not
chip or tear the surface of the stock when running against
the grain on the last half of each piece. Both feed rolls
are arranged to bear on the edge of the stock so that a continuous power. feed is maintained.

When running heavy work over circular forms it is frequently necessary to remove one or both top feed rolls and feed the stock by hand, especially if the circle is of small radius.   One must always be guided in such emergencies by his own good judgment as no hard and fast rules are applicable.

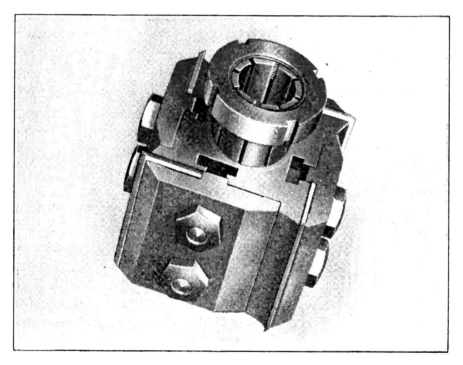

A square, self-centering side head fitted with thin steel knives which are clamped in place with caps and bolts.

CHAPTER VII.

RUNNING MOLDING FACE DOWN.

Early types of molding machines were designed to make molding face up, evidently because this seemed to be the only proper way to run it. Bottom heads, therefore, were made smaller than top heads, and driven by lighter belts and smaller pulleys. The overhanging part of the machine bed which carried the bottom head was none too well supported. In those days probably no one thought of running molding upside down, and if he had attempted to do it with the machines then ·in use he would more than likely have made a complete failure. The bottom heads were too light and did not have sufficient belt power; there was not enough room to accommodate the swing of large knives, and being mounted at the extreme end of a long overhanging bed, the bottom head would have produced wave marks on the face of the molding if heavy cutting had been attempted. Some primitive types of molders are still in service, and needless to say, it is out of the question to try to do anything but very light work with the bottom heads.

Modern molding machines, however, are built differently. The bottom heads are equal to the top heads in size and power, and the space around them is adjustable to permit the swing of reasonably long knives. They are amply supported in a massive machine frame. The bed plate at the rear swings down instead of sideways, thus giving convenient access to the knives. It is an easy matter to run molding face down on the later machines; in fact, practical moldermen have learned that in many cases much better results are obtained by practicing the face-down method.

Numerous arguments abound in favor of running ordinary molding face down, but probably the chief reason responsible for a general adoption of the face-down practice is the increased cost of lumber and the consequent tendency of saw mills to saw more closely, scanting the lumber generally instead of sawing full thickness. Shrinkage, due to kiln-drying, leaves the stock even thinner, hence when it is ready to be manufactured into molding a large percent of it is likely to be considerably scant of standard thickness.

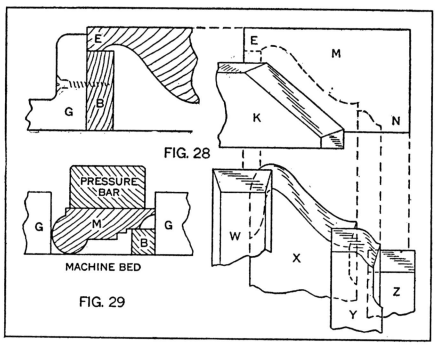

Fig. 28. K is roughing cutter on side head. W, X, Y, Z are combination of cutters for working mold M. B is wood block which supports overhanging edge E. Fig. 29. M is picture mold being worked face down. B is supporting block.

For example, suppose 4/4 stock is to be used for molding which finishes 25/32-in.; if run face up it must measure about 15/16-in. in thickness or rough spots will appear on the face of the molding; if run face down, however, it can be scant 7/8-in. thick and still make perfect molding. The face-down method, therefore, is often the means of preventing the loss of an immense amount of good lumber on account of the scant thickness.

Another decided advantage in the face-down process,

which is especially noticeable when making molding from unsurfaced lumber is that the bottom head almost invariably produces smoother work than the top head. This is due to the fact that the bottom head always takes a cut of uniform depth regardless of the varying thickness of the rough sawed stock, and because it is far removed from the infeed rolls and top chipbreaker. The bottom head also produces more accurate work as a rule, because when the stock reaches it the material is leveled off and sized to uniform thickness and width, and closely confined between side guides and under the pressure bar.

When moldings are run face down, practically all formed pressure bars are dispensed with. Only a few plain flat bars of different widths to suit the various widths of molding are needed. If a small chip or part of a knot lodges under the pressure bar and scores the molding, no harm is done because the damage is on the back, whereas, if the molding is run face up the least scratch on the top surface may ruin it. It often happens that several hundred feet of molding are run before the operator discovers that the surface is being scored and, of course, if it is being worked face up this amount is spoiled.

There is an additional convenience in setting up a bottom head for new molding and checking the correctness of the set-up because, when the pressure bar is set down properly, one can feed the first piece a few inches over the bottom head and quickly determine its degree of correctness without removing it from the machine. If edge-cutters or splitters, which cut entirely through the molding are employed, they can be used to greater advantage on the bottom head than on the top head because there is no danger of cutting into iron if they are set out a fraction of an inch too far.

Here is one more point worthy of mention. When feeding anything but the very smallest moldings, if the last end of a piece passes the infeed rolls without another closely butted against it, an ugly mark is usually made across its top surface by the top head. This end is therefore spoiled if the face side of the molding is at the top, but if it is on the bottom, no damage is done.

In Fig. 28 there is illustrated another advantage in the face-down method which may be taken into account when any molding requires extra heavy cuts along one or both face corners. These cuts are often difficult and dangerous to make with single knives on one head, but by running the material face down, the heavy cuts can be divided between knives on the side and bottom head. For instance, in Fig. 28, a plain bevel knife K on the side head cuts away more than half of the surplus wood and leaves only a comparatively light cut for the finishing knives W, X, Y, Z on the bottom head. This is a more safe and easy way to run large moldings of this type and it is a method which invariably produces a smoother-finished product. There is so much of the under side of this molding cut away, however, that only a very narrow part N rests on the bed as it leaves the bottom head. Therefore, the last end of each piece is likely to "cave into the knives" as the cut is completed. This is prevented by a block B which acts as a support for the overhanging edge E. In fact, suitable blocking should always be used, as illustrated at B, Fig. 28, and B, Fig. 29, to aid in supporting moldings which are under-cut to such an extent that they are apt to "cave in" or "roll" as they leave the bottom head. The adjusting of what blocking is required does not consume much time—not nearly so much as would otherwise be taken up in making, keeping in order and adjusting formed pressure bars for top-head work. The blocks are short and easily attached temporarily to the rear bed or side guides by wood screws.

There are, of course, some moldings which cannot be run face down and others which, altho they can be run face down, work to better advantage face up. In doubtful cases one must use his best judgment in choosing the safest, easiest and most practical method.

CHAPTER VIII.

SPECIAL SURFACING AND MILLING KNIVES.

When comparatively straight-grained lumber is manufactured into moldings, special knives and other devices are not required. But in working curly or cross-grained stock, or in making a groove or rabbett without planing off the flat surfaces adjoining it, there is need for something more than regular cutters and ordinary methods of knife setting to produce smooth work.

One of the most general methods employed to obtain a smooth finish when working curly or cross-grained wood is to back bevel the knives as shown at B, Fig. 30. A knife ground in this manner makes a scraping, non-tearing cut instead of "picking up" and tearing out the grain. More power is required, however, to drive it thru the wood and the method can only be used successfully on thoroly kiln-dried lumber. Equally as good results can be obtained by reversing the knives on the head (the knives being turned upside down) but this is only recommeded in emerg·encies and on short runs because knives running in this manner are necessarily limited to light cuts on account of the added power required to drive them thru the stock.

Another method of preventing torn grain in flat work is to set ordinary knives so they have about half the usual projection past the lip of the head, say a scant 1/16-in. if that is sufficient to give clearance for the bolt heads. Should there be insufficient clearance when the knives are set so close, or, if the lip of the head is nicked and in bad con·dition, the same effect can be obtained by placing a piece of thin, flat steel (part of a resaw blade will do) under the knives as shown at C, Fig. 30. The piece of steel must be slotted like the knife and its working edge ground perfectly straight. This edge is set back only 3/64-in. or

cuts the face of the rabbett, see S to T, Fig. 33, and a special knife like Fig. 34 or 35 cuts the edge of the rabbett as E to S, Fig. 33. The knife shown in Fig. 34 is really an ordinary beveled knife turned upside down on the head

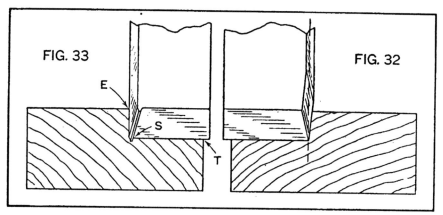

Figs. 32 and 33. Methods of fitting up rabbetting knives to make non-tearing cut.

but the side clearance is reversed. Use is made of a side spur, S A, and only a very slight side clearance C, just enough to prevent the back of the knife from striking the edge of the rabbett. The knife is set so spur, S A, cuts precisely the depth of the rabbett. One might think that the point S of the spur does the entire edge cutting but this is not the case. The cutting edge extends from point S to point A, Fig. 34, and it produces a non-tearing, slicing cut. The knife shown in Fig. 35, with its knife-like curved edge, is forged to shape and ground to make a slicing cut similar to that just described. It is shown in the form of a spike bit, but can be made slotted if desired. Great care is required in setting either of these special edge-cutting knives, but once a knife of either type is properly sharpened and set, it will cut a perfectly smooth square edge without tearing, regardless of the grain or kind of wood being worked.

Knives for cutting grooves should be made and used in pairs of rights and lefts, that is, one to cut one side of the groove and its mate to cut the other side, so that as the side edges of the knives wear away by repeated sharpening they can still be set out to cut the standard width grooves for which they are intended. Instead of using grooving knives with side spurs, as shown in Fig. 33, one can, if he prefers, obtain equally as good results with knives which have a shear-cutting bevel on the front. The knives are paired and positioned on the head so the sharp, shear-cutting edge is turned to the outside to make the edge cut.

In Fig. 36 is shown an excellent cutter for milling narrow grooves. There should be a pair of these cutters, and

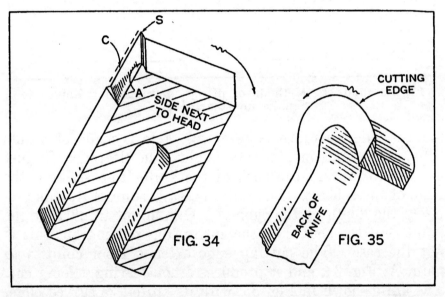

Figs. 34 and 35. Special knives for cutting the edge of rabbetts.

the saw teeth instead of being filed square across at the front should be given a slight lead or bevel toward the outer edge, which is to the left for one cutter and to the right for the other. This produces a double shear cut. The front teeth of each cutter are lower than the back teeth so that they will not do most of the cutting when a fast rate of feed is carried. Cutters of this type stay sharp a long time and cut a remarkably smooth clean groove in any kind of wood at fast feed. They will even produce a smooth groove on surfaced stock without the usual "skinning off" of the face side. In fact, the first cutters of this type ever made were designed to meet just such an emergency. The cutters are made of ordinary slotted steel blanks which are first cut at A, Fig. 36, then heated a cherry red and bent

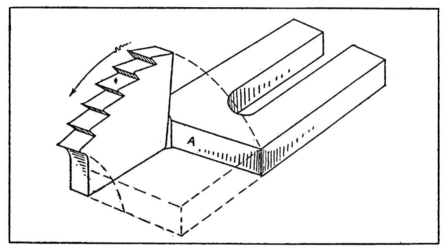

Fig. 36. Special grooving cutter.

as shown. The bent upright part of the cutter is scribed and roughly ground to the cutting circle of the head, after which the teeth are formed, jointed and sharpened.

Grooves and rabbetts are often cut with thick saws, built-up thin saws, wabble saws, Shimer heads, and various high-speed cutterheads mounted on the proper spindles of the molder, and while these are all excellent methods they are in most instances only profitable on long runs of standard patterns. In many cases some of these latter methods can be used with good results for making various irregular-shaped moldings when unusually cross or curly-grained stock is encountered.

CHAPTER IX.

BRACES AND KNIVES FOR HEAVY WORK.

When making large moldings which require deep heavy cuts with long knives, the need of braces or reinforcing devices, and sometimes special knives, immediately becomes apparent. The use of extra large knives is dreaded by some moldermen because of the possible danger involved, but if the knives are properly made, accurately balanced, bolted and braced in a substantial manner, there is scarcely any danger of having a "smash up." By properly made, it is meant that they should be made of good steel, somewhat thicker than the ordinary knife steel, and that the cutting edges should only be hardened a short distance back, thus leaving the main body of the knife tough rather than hard and brittle. A very hard knife is almost sure to break in a deep or heavy cut no matter how well it is braced.

In designing or selecting a knife brace one must always bear in mind that the chief object of bracing is to prevent the overhanging end of the knife from bending back or shearing its bolts. Another thing to consider when selecting braces and planning a set-up for heavy work is the amount of molding to be run, whether it is a standard pattern or just a special job which may never be called for again. It is neither wise nor profitable to spend a great deal of time fitting up something elaborate for one small job, and it is equally unwise and often dangerous to fit up a temporary makeshift for running stock patterns which are repeatedly ordered in comparatively large quantities. Practically every job of heavy molding presents problems peculiar to itself, but with the assistance of the following illustrations of devices and descriptions of their uses, any molderman should be able to quickly solve the difficulties that may arise and turn out the most complex kinds of molding with ease and safety.

Often times the only kind of reinforcement a knife requires is an additional bolt at each side. For example, if a single slotted knife, say 3-in. or more in width, is set to take a medium heavy cut the strain of cutting is likely to be more than one bolt can hold, but, if a bolt is added to each side as shown at A, Fig. 37, thus making three in all, the knife can be safely held in place. The holding power

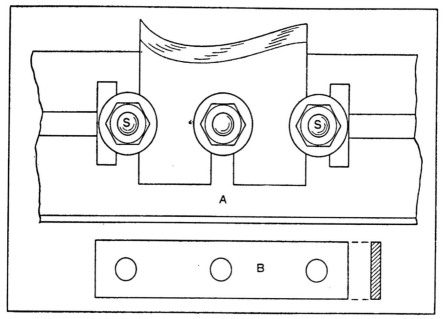

Fig. 37.   Reinforcing a wide knife with bolt at each side.

of the side bolts S, S, Fig. 37, can be further strengthened if a very short slot is made in each side of the knife to receive about half or three-quarters of each bolt. Another good practice is to remove the bolt washers and substitute in their stead a soft steel plate as pictured at B, Fig. 37. This plate is drilled to receive the three bolts and when in position it spans the entire knife, thereby serving the dual purpose of clamping cap and bolt washer.

Long, overhanging knives require bracing at points beyond the lip of the head and reasonably close to their cutting edges. Some very effective braces for long knives are shown at C, D, E and F, Fig. 38, and at G and H, Fig. 39. The braces C, D, E and F, are simply different types of side braces, all being made of steel. Each brace has a mate to balance it. It is also a good plan to have right and left

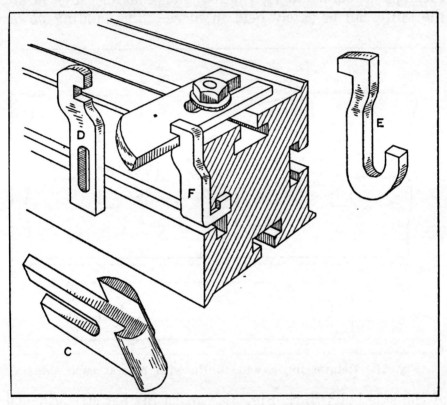

Fig.38. Four different kinds of side braces.

side braces so that knives can be braced on both sides if necessary. The brace C is made of ordinary slotted steel and arranged to hook over the lip of the head and rest in position alongside the knife it reinforces. Braces D and E are also designed to hook over the side of a knife. They are bolted to the side of the head immediately in front of the knife which they brace. The brace F is similar to D and E except that it is bent around at one end to fit into

the bolt slot as shown and therefore does not require any
bolts. The three braces D, E and F, are each offset, as
illustrated, to clear the lip of the head and bring the
"anchor point" near the cutting edge of the knife.

When the shape of a long knife is such that it cannot be
braced along the sides with braces as shown in Fig. 38, an

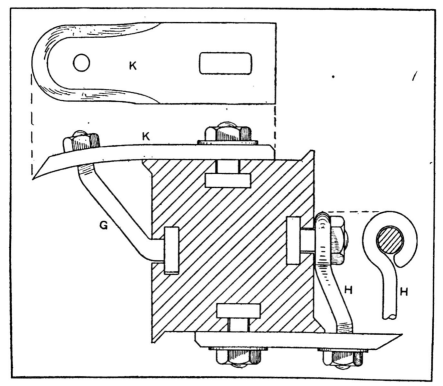

Fig. 39. Two ways to brace extra long knives.

effective method of anchoring it is to drill a hole near the
middle of the projecting part and insert a long bent bolt
as indicated at G, Fig. 39, or an eye bolt as shown at H,
Fig. 39. The type of anchor bolt chosen should always be
used with a mate of the same kind, size and weight in order
to make a good running balance when the machine is set
up.

As a further safety precaution it is well to have closed

slots in all extra long knives, see K, Fig. 39. A knife with a closed slot has greater strength at the back where it is bolted to the head. It cannot spread at the slot nor get away during service without stripping its bolts. These extra long knives should also be bent slightly forward, as

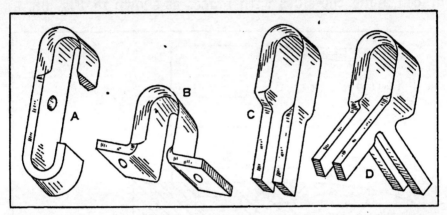

Fig. 40.   Four types of "scoop" or "loop" knives for making gutter.

shown at K, to give the cutting edge a more acute cutting angle because, when bent forward, they cut more easily, require less power, and consequently cause less strain on the bolts and braces.

While the type of knife illustrated at K, Fig. 39, will serve fairly well for making deep cuts, the so-called scoop or loop knives (four different kinds of which are pictured in Fig. 40) are far superior in every way. The cutting edges of knives A, B, C and D, Fig. 40, are the same but the manner in which the different types of knives are bolted to the head differs. These knives are made of good bar steel and their cutting parts are bent to make the shape desired. Proper allowance is naturally made for clearance in each case. Knives of this kind will cut thru solid wood with remarkable ease and produce real smooth work. The shavings and air pass right thru the loop of the knives, hence there is much less resistance offered to their motion and less power is required to drive them. This means an added element of safety which is important.

The knife A, Fig. 40, has two cutting edges and is designed to slip onto a cutterhead spindle. A pair of the knives should be used, one with the ends bent to the right,

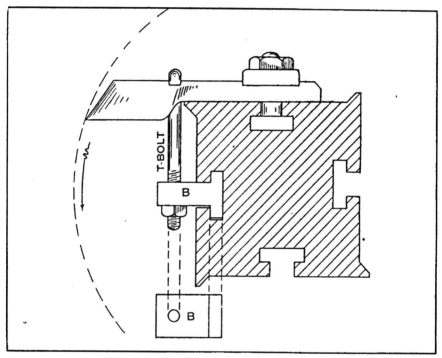

Fig. 41. Bracing a loop knife with T-bolt and steel block B.

the other with ends bent to the left, and the cutters should be separated by spacing-collars and clamped in position on a spindle in the same manner that a saw is fastened to its arbor. One thing is certain about knives of this type, they cannot get away after being properly clamped onto a spindle. This is a point in their favor but, on the other hand, when they are used no other knife can be positioned with them on account of the absence of a cutterhead.

Knives B, C and D, are made in pairs and designed to bolt directly onto any ordinary square-slotted head. Knife

C is made of bar steel which is thicker but narrower than the steel used for making types A, B and D. The cutting part is heated and hammered thinner and, wider when the

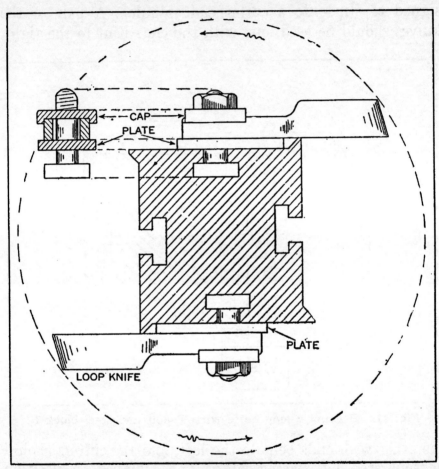

Fig. 42. Another method of clamping loop knives to a square cutterhead.

knife is made. Knife C possesses a marked advantage over the other three types of scoop knives because it is adjustable for depth of cut. Fig. 41 shows one method of bolting and bracing a knife like C, Fig. 40, but a better method appears illustrated in Fig. 42. Usually the lips on two corners of the head are filed down level with the sides of the head to accommodate a pair of knives that are put on in this way, but it is not absolutely necessary to do this since the two sides of the head can be blocked up with plates, as shown, to make them come level with the edges of the lips. This method of positioning the knives gives them

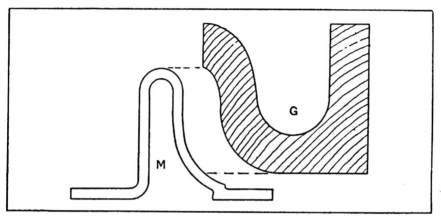

Fig. 43. G is a pattern of ogee gutter. M is loop knife for cutting the ogee.

solid backing all along the back and up to their cutting edge without extra bracing. A single powerful bolt and square cap which fits over the knife, as illustrated, are usually all that are needed to hold each knife on the head. This method of putting knives on a square head gives the bolts greater leverage, and as a result the knives are easier to hold and the element of danger greatly reduced.

Knife D, Fig. 40, is designed to straddle the corner of a square head and is bolted on both sides, therefore, it requires no extra bracing. In making any of the knives shown in Fig. 40, good steel of generous thickness should be used in order to offset any danger of the knives collapsing while in service. The steel for knives A, B and D, should be 3/8-in. or 7/16-in. thick while that for knife C should be about 9/16-in. or 5/8-in. in thickness if the knives are intended for making cuts 2-in. or more in depth. Knives of the type just described are used quite extensively in Pacific Coast mills for cutting different patterns of heavy solid wood gutter, one form of which is shown at G, Fig. 43. They are also used for various other kinds of heavy cutting.

Really the best way to make plain deep cuts such as gutter, trunking channels, deep grooves, rabbets, etc., is to use sectional slip-on cutterheads. When this equipment is available an enlarged section or disc of the required width is positioned in line with the deep cut, and the remaining part of the molder spindle simply carries sections of normal size cutterheads, either square or round. This permits the use of ordinary small knives set at a normal projection, therefore braces are unnecessary and virtually all danger is eliminated.

A somewhat crude modification of this idea consists of using steel blocks which are bolted to the sides or fitted over

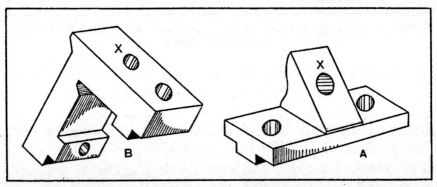

Fig. 44.   Two kinds of blocks for enlarging square cutterheads.

the corners of ordinary square slotted heads.. Two such blocks are shown at A and B, Fig. 44. It will be noticed that block A is designed to be bolted to the flat side of a square head while block B is made to fit over the corner of a square head. Both blocks contain bolt holes running thru the projecting parts which fit into the head slots. The blocks also contain holes X, X, which are tapped for planer bolts to hold the knives. Ordinary knives bolted to these blocks require only a normal projection to make very deep cuts.

Sometimes it is necessary, or at least desirable, to run material that finishes an inch or two wider than the rated capacity of the molder. For instance, there may be an order for 10-in. base and the only machine available is an 8-in. molder. This means that the surfacing knives must extend out over the end of the head 1½ to 2-in. Fig. 45

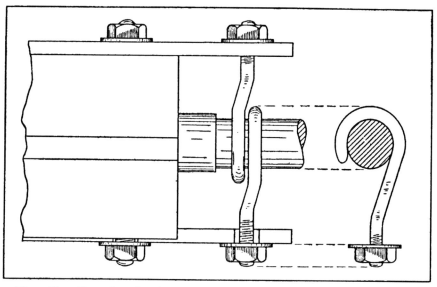

Fig. 45. Showing hooks in position for holding ends of projecting surfacer knives.

shows how the overhanging ends may be supported or braced with a pair of iron or soft steel hooks. This is a make-shift method but it will serve the purpose very well in emergencies provided a comparatively light cut is taken at a moderate rate of speed. If the outside head does not pull out far enough to clear the work it can be fitted with a special small head or pair of shaper collars and shaper knives to make the edge cut. Otherwise the side head can be removed from the spindle and the outside edge of the wide molding finished afterward on a jointer or shaper.

CHAPTER X.

MAKING MOLDINGS IN MULTIPLES.

In manufacturing large quantities of narrow moldings, the cost of production can often be materially reduced by working the moldings in multiples or gangs of two, three, or more at a time instead of ripping stock into narrow strips and running them singly. Of course, there should be enough molding of a kind to justify the extra time and expense incident to making the more complicated set-up. Usually it is not advisable to run anything except stock patterns in multiples and even they should not be run in gangs from good wide lumber if there is plenty of narrow material on hand for making them singly. The extra cost of wide lumber must always be taken into consideration when planning on converting it into narrow moldings. There is certainly no satisfaction in making an imposing display of higher molder efficiency by running molding in gangs if, after the job is completed and properly figured, it is found that the finished moldings are worth less than the market price of the wide lumber from which they were made.

Manufacturers of wholesale softwood molding, picture frame and embossed molding, run most of their narrow patterns in multiples of two or more. There is this difference in the established methods of making the different classes of moldings. In planing mills where large quantities of woodwork for building purposes are manufactured, multiple work is generally run face down, while in furniture and picture frame factories almost everything is made face up. Any ordinary pattern of molding can usually be run successfully either way altho there are some practical advantages in the face-down system under certain conditions, as mentioned in Chapter VII. The best way to run any multiple molding depends largely upon the profile of the molding and the manner in which it is most practical to separate the multiples. For instance, if the back of the molding is flat, it can be run face down very satisfactory

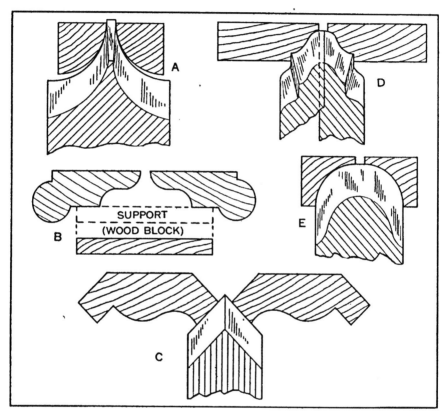

Fig. 46. Examples of moldings made in pairs, face down.

by surfacing the back with the top head, the square edges with the side heads, and finally molding and splitting it, as required, with the bottom head, see examples in Fig. 46. When this method is followed no wide stock is separated into narrow pieces by the top head, hence there are no narrow strips to twist, break, buckle nor get out of line in the machine. Here is another point, when the knives which separate the moldings are on the bottom head, they do not require such fine depth adjustment as when they are positioned on the top head. If the knives which divide the moldings cut 1/16-in. or ⅛-in. deeper than necessary, it is all right; the points simply cut a little groove in the underside of the wood pressure bar and no harm is done. On the other hand, this extra length is never permissible on the top head unless the bed plate directly under the head is recessed especially to accommodate the swing of long knife points.

In Fig. 46, five different moldings are illustrated in the relative position in which they leave a molder when made

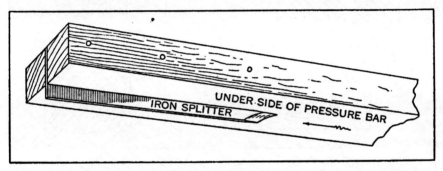

Fig. 47. Splitter or center guide in rear end of pressure bar.

in twos. They are run in pairs of rights and lefts as a matter of convenience. Patterns A, B, and C are separated by the points of double-edged molding knives which cut thru the wood at the dividing line. In order to separate patterns like D and E, a narrow straight knife or some kind of splitting cutter must be used to cut thru the square inside edges. For the purpose of holding the multiples apart as they leave the bottom head, after being separated, a splitter is fastened in the pressure bar at a point just back of where the knives strike, see Fig. 47. This splitter serves as a sort of center guide and prevents the moldings from shifting or playing sidewise, and thus becoming marred at the ends by the bottom head knives. It is not always necessary to use a splitter in the pressure bar to separate multiples but in the majority of cases it is very advantageous. The question of when and where to use a splitter, if at all, is another one of those things which the molderman must decide by the exercise of good judgment based upon practical experience.

Where the profile of the molding is such that there is an overhanging, unsupported edge left when the bottom head knives complete their cutting, wood blocking must be fast-

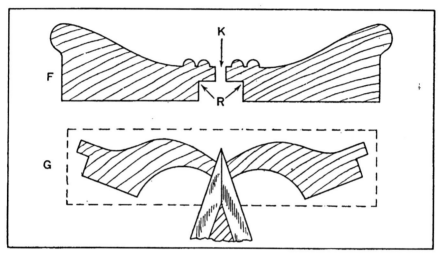

Fig. 48. Picture frame moldings made in pairs, face up.

ened to the rear table, see illustration B, Fig. 46, to prevent the last ends from caving into the knives as the molding leaves the machine. See other examples in Chapter VII.

Referring to Fig. 48, two patterns of picture frame molding are shown at F and G as they appear leaving the machine when run face up. A great many picture frame moldings are run in twos in a manner similar to that illustrated at F, Fig. 48. The top profile and the dividing kerf K are cut by knives on the top head while the rabbett R is made by knives on the bottom head. The moldings are not separated therefore until the rabbett R is cut with the bottom head. Frequently this rabbett is not made on the molder at all because it is not desired to separate the moldings until after they have been passed thru an embossing machine and perhaps a molding sander. The rabbett is finally cut and the moldings separated by passing the double molding thru a machine fitted with solid rubber feed rolls and special saws or cutters for milling a wide central groove which, when completed, leaves an edge rabbett R on each piece of molding.

SPLITTING CUTTERS.

Among the many splitting cutters which have been devised from time to time by moldermen and tool makers all over the country, the few which are standing up best and proving most practical in all respects are illustrated in Figs. 49 and 50. Referring to Fig. 49, cutters A and C are designed to be fastened with machine screws on the side of a block like B, which is bolted in the slot of a square cutter-head. Cutter A is part of a saw blade and its teeth should be swaged for best results. Cutter C is high-speed steel ground thin at the back and near the bottom for clearance, and is beveled 45 degrees on the front or cutting edge. Its mate, which belongs on the opposite side of the head is beveled in the opposite direction to equalize or balance the cut. Block B, fitting in the slot of a square head as it does, serves to hold its cutter perfectly in line. The cutter D is made from a section of saw blade and arranged to fit into a narrow slot in a steel knife blank E, which serves as its holder. The cutter is secured rigidly in place by a machine screw as shown. The cutter at F, Fig. 49, is made from a single piece of steel $\frac{1}{4}$-in. or 5/16-in. thick. The steel is hammered thin, while red hot, and bent and shaped so the front or cutting edge is thicker than the back to give clearance. The front concave cutting edge is ground hollow with a thin emery wheel so that both edges are cutting edges instead of only one as at C, Fig. 49.

In Fig. 50, there appears a splitting cutter, a little more elaborate in design than those just described. The holder is made of soft steel and is in two parts, G and J, which are hinged at the back to permit opening the holder and inserting a high-speed steel cutter H. This holder folds up

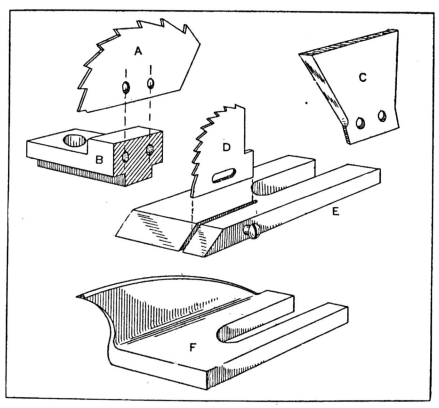

Fig. 49. Four types of practical splitting cutters.

over the cutter in a very compact manner and does not take any more room on the head than an ordinary narrow slotted knife. The whole device can be attached to any square head with a knife bolt. The cutter in this device, like those illustrated at A, C, and D, Fig. 49, is removable and renewable.

There is quite a variety of uses which splitting cutters

can be put to in molder work. They serve not only as splitting cutters for ripping square-edge stock and separating multiple moldings but are also used for creasing or kerfing the backs of thick jambs, etc., and for cutting the square edges of deep rabbetts and grooves. The chief re-

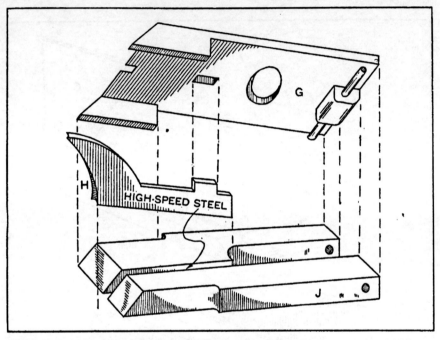

Fig. 50. Special high-speed steel splitting cutter and its holder.

quirements of a successful splitting cutter are plenty of back strength and sufficient clearance. The cutter must also be positioned square with the head, perfectly in line with the travel of the stock, and needless to say, it must be secured in a rigid manner so there will be no chattering or vibration when in action.

Saw tooth splitting cutters like A and D, Fig. 49, should be of such shape that the forward or first teeth are proportionately lower than those at the middle and back, in order that the cutting will be evenly distributed among all the teeth. Otherwise, a few forward teeth of each splitter will carry most of the strain and do most of the cutting, with the result that the cutter will heat rapidly and not stand up to heavy work at fast feed.

While the examples of multiple work appearing in Figs. 46 and 48 show only simple moldings in pairs, each member of which is the same size and shape, it does not follow that all moldings run in multiples are or must be made in this manner. There is really no limit to the variety of combinations that can be worked out. Entirely different pat-

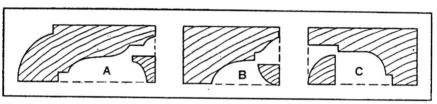

Fig. 51. Showing how an extra molding can sometimes be saved by under-cutting.

terns can be paired if desired and, instead of running moldings in twos, the machine can be set up for making three, four, five or more strips simultaneously from one piece of stock. So far as the mechanical part of the operation is concerned, one might run 12-in. stock in gangs of narrow moldings right along, but considered from a business standpoint, this practice is rank folly on account of the high cost of wide lumber.

Sometimes thin moldings are run in gangs but in double thicknesses, and they are put thru a band resaw afterward to separate them. This practice is occasionally followed in the manufacture of screen moldings and other cheap work. Other patterns are run double thickness and split apart with a circular saw attachment on the rear of the molder.

SAVING AN EXTRA MOLDING.

When the profile of any single molding 13/16-in. or more in thickness is such that one or both face corners must be cut away to a considerable extent, it is sometimes possible to save a small molding or strip by doing a little under-cutting as shown at A, B, and C, Fig. 51. This method of making moldings, however, is not practiced very extensively and it is not recommended excepting on long runs of softwood molding when good lumber, practically free from knots, is worked. The method is entirely practical and is the means of producing or rather saving an extra strip of small molding which would otherwise be cut into shavings. If the principal molding of the cluster is made face down as shown in the examples in Fig. 51, it is generally advisable and often necessary to fasten wood block-

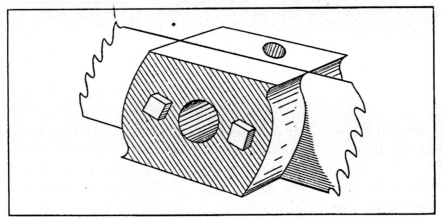

Fig. 52. Combination head for splitting and planing.

ing onto the rear guides and back table in order to hold the moldings solidly in place and prevent them from caving in to the knives as they leave the bottom head.

One more method of gaining an extra strip of molding without the use of additional stock is to saw out the rabbett of rabbetted patterns, such as jambs, screen door stock, etc., instead of cutting it into shavings, with knives. The strips are sawed out with either circular saws, splitting cutters, or special cutterheads in which square or round-head sections carrying knives are combined with circular saws, or parts of saws. One type of special head for this purpose appears

in Fig. 52. This is used on the side spindle to surface the
edge and saw under the strip which is saved. The practice
of saving strips of molding, as described, is followed quite
extensively in factories making screen doors, bee hives, and
other stock products which are manufactured in large quantities. In Fig. 53, there is shown the manner in which a

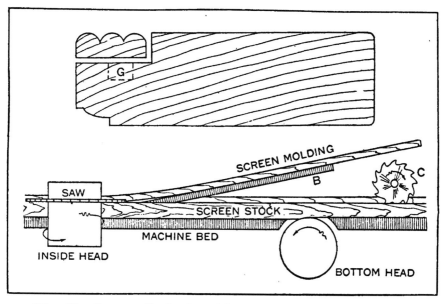

Fig. 53. One method of making screen door stock and saving the molding. Fig. 54. Machine set-up for making molding
in Fig. 53.

strip of screen molding is sawed from the corner of screen
door stock as the material feeds thru the machine. This
operation in itself is simple indeed, but when the specifica-
tions call for a small groove G in the bottom of the finished rabbett the proposition is not quite so easy. This
groove, by the way, is milled for the purpose of receiving the
wire and a small strip of wood or rattan which crimps the
screen wire securely in place. The usual method of milling
the groove without resorting to an extra operation and an
extra handling appears in Fig. 54. A small diameter grooving saw C is suitably mounted directly above the rear

table and driven by a separate countershaft, a motor, or by a short belt from a narrow pulley alongside the bottom head pulley. The strip of screen molding is sprung upward, after being cut free from the stock by a saw or cutter on the inside head. As the strip advances over the inclined bar B, it clears cutter C, and when finished it either falls into a trough or is taken away by the helper. The cutter can be rotated in either direction because it makes a very light cut. The yoke which carries the small grooving saw is removable so that it can be detached after a job of this kind is completed.

Gang of splitting saws mounted on self-centering sleeve for use on molder spindle. The saws are separated by spacing collars.

CHAPTER XI.

MISCELLANEOUS MOLDER WORK.

There is a vast amount of special work produced on molders in various kinds of wood-working factories and many freak jobs are occasionally done in jobbing mills in emergencies, but since the scope of each of these individual

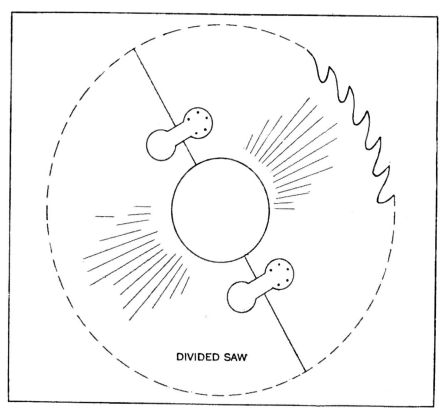

Fig. 55. Divided rip saw for use on top or bottom spindle.

jobs is so limited it is hardly worth describing all of them in detail. For example, there are cases where rope or twist molding has been successfully made by turning round molding spirally by hand in a form clamped diagonally over the bottom head of the molder. An occasion seldom arises, however, for doing such unusual work in this crude manner. Twist molding of practically any design and pattern is now made in factories where regular twist machines are in operation. A few examples of special work which are perhaps of more general interest follow:

RIPPING WITH DIVIDED SAWS IN GANGS.

Divided rip saws like Fig. 55, are sometimes used between large collars on the bottom spindle of a molder for ripping lattice, parquetry strips, and other light work. The strips are first planed with the top head and then ripped in the same operation. Fine-toothed, hollow-ground saws are best for this work because they cut smoothly and are easier to fit and keep in order. Plain spring set, or swaged saws can be used with good results, however, if they are carefully fitted. The halves are held together with keys which are fastened on one side, as shown. Being divided, the saws can be put on and removed at any time without disturbing the boxes.

RUNNING HEAVY MOLDED CASKET SIDES.

Wide casket sides, having a narrow piece glued on the face side next to the bottom edge to make the base, and,

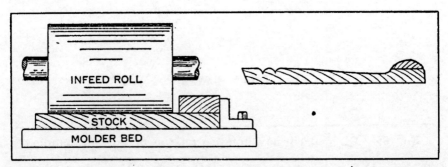

Fig. 56. Position of top rolls for feeding special casket sides.

sometimes a piece at the top to make a heavy ledge, are run face up with the thick edge next to the guide rail.

Altho full-width top feed rolls can be used for this work, better results are obtained by the use of narrow rolls which ride only on the thin part, as shown in Fig. 56. A sectional chipbreaker is also used so the thin part of the sides can be held down firmly to the machine bed as the sides advance to the top head. In running wide material, the under-side of the pressure bar should be grooved, recessed, or cut away at all points except where pressure is absolutely needed. This precaution serves to reduce excessive friction between the molding and the pressure bar. Otherwise the material will not feed freely, but will stick in the machine or tend to "crawl" away from the guide rail.

MAKING GLUE JOINTS.

Edge glue joints can be made on a molder on condition that the stock is fairly straight and of uniform width. In one casket factory the 1x10 common cedar for adult-size

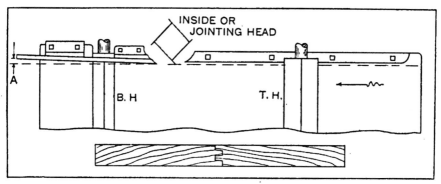

Fig. 57. Line-up of guide rail and inside head for making slack-center glue joints.

casket bottoms is successfully edge jointed on a molder. The stock is first roughly cut to about 6-ft. in length and then sent to a 15-in. molder where the pieces are surfaced two sides, jointed and sized to exact width in one operation. A reversable tongue and groove joint is worked on one edge with the inside head which is set to take a full ¼-in. cut. The stock is of such size that it does not spring out of true in the machine provided the latter is in proper alignment and adjusted to feed freely. Accurate setting of the jointing (inside) head, guide rail, and outside guides is particularly important. In actual practice the operator employs a little trick to make the joints slack or slightly concave in the center to offset the possibility of some joints being full in the center. The trick lies in the adjustment of the guide rail back of the jointing head, and it is the same scheme that is used to make slack center joints on an ordinary glue jointer. A continuous back guide rail is used, and at the rear of the machine it is clamped slightly out of line as shown exaggerated at A, Fig. 57. This method of jointing, planing, and sizing material to width in one operation effects considerable saving in manufacturing costs and the idea can often be applied to other kinds of work with good results.

RUNNING VERY THIN WORK.

In order to successfully run real thin patterns, one must, as a rule, resort to a method similar to that used in planing thin veneer, that is, run it on a hardwood board. The board is fed thru the machine with the thin material on top in practically the same way that a form is used. Material can be planed as thin as 1/32-in. in this way because the boards support it as it passes thru the machine.

MILLING A TAPERED CHANNEL IN WOOD DRAINS.

A molder on which the bed is stationary, and the top head and rolls are adjustable vertically, can be used for quite a variety of special work. A tapered channel can be milled in solid wood, as the stock feeds thru the machine, by gradually cranking the head to make a pointer follow a line scribed on the outside to correspond to the required taper. Square pilaster sides can also be fluted up to a point near each end by dropping the top head to cut at the proper point and then raising it when the material has fed forward a distance equal to the required length of the flutes.

MOLDING ACROSS THE GRAIN.

In piano, furniture and novelty plants, certain patterns must be molded directly across side and end grain. Piano actions serve as a good example of work which must be molded smoothly and accurately crosswise instead of lengthwise of the grain. Altho ordinary narrow knives are sometimes used in this class of work, knives which are slightly twisted or ground, and positioned so that they make

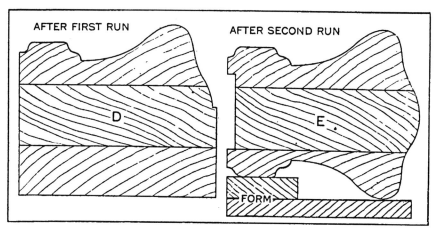

Fig. 58.　Method of making stair rail in two runs.

a shearing cut like tenoner and automatic turning machine knives, are the most satisfactory because they do not pick up nor tear out the grain. Circular milled cutters, with an angular face-bevel instead of the usual straight face, are also suitable for molding across the grain, especially when they are accurately ground so that all wings do an equal share of cutting. Usually the material to be run crosswise of the grain is planed and sawed into blocks which are glued up in lengths suitable for feeding thru a molder. In many cases the regular chipbreaker is dispensed with and hardwood springs are set right up against the cutting knives at the point where they begin cutting, while the point of a wood pressure bar is set as close as possible to the point where the molding leaves the cutter-head. The object in setting the chipbreaker and pressure bar so close is to reduce the open gap for the cutters to a minimum and thereby prevent chipping and tearing of grain. It is the same principle that is employed in shaper work.

RUNNING STAIR RAIL.

A good rule to follow in making stair rail is to place the greatest width flatwise and the heaviest cut to the top head. This is why so many patterns of stair rail are run on their sides instead of straight up. In the latter case, the bottom of the rail is turned to the guide rail. When large quantities of standard stair rail are made, the work is often performed in one operation, but when there is only a small amount to make, or if the bottom head is too light to make the finish cut, the rail is made in two operations. On short runs two operations are often really more economical than one. For instance, suppose a rail is the same shape on both sides. One set of knives on the top head will serve for both runs without any change except to adjust the head for thickness. By making the rail in two operations the set-up time alone is reduced to less than half. If the pattern is such that new knives are required a further saving is naturally effected by working it in two operations. In Fig. 58 is shown a simple stair rail and the method of making it in two operations. After the first run the rail appears like D, Fig. 58, then it is turned end for end and run on a light hardwood form as shown at E, Fig. 58. During the second run the bottom of the rail is grooved for balusters with the inside head.

MAKING CHURCH SEATING.

On account of its great width and depth of cut, church seating is generally made on a heavy 18-in. molder. While ordinary knives and heads can be used for molding church seating, a special head with detachable, close-fitting formed lips serves the purpose much better because it effectually prevents all chipping and torn grain. Heads of this type, of course, are only recommended for factories that produce large quantities of church seating or other standard work. The four corners of the head are milled out to receive four formed chipbreakers like 'M, Fig. 59, which are ground to

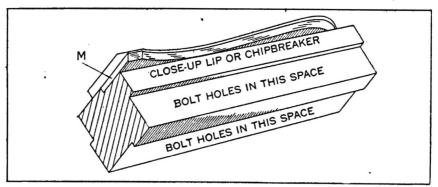

Fig. 59.   Special head, rabbetted to receive formed lips like M.

match the molding knives. These formed lips or chip-breakers are really reversed-knives without slots and are screwed down to the head with flat-head machine screws. The cutting knives are bolted firmly with planer bolts threaded down into the solid head. The cutting edge of the knives is set out only about 1/16-in. beyond the corresponding edge of the formed lips. The result is very smooth work, no torn grain, and less strain on the knives. The formed lips for church seating are detachable and can be replaced with straight steel lips or lips of some other shape, if so desired.

CHAPTER XII

HIGH-SPEED MOLDER WORK.

The modern high-speed molder, with massive frame, large journals, wide pulleys, improved feed mechanism and multiple bit, self-centering slip-on heads carrying high-speed steel cutters, represents the acme of perfection in the development of wood-molding machines. Like all other fast-feed machines, it is designed to meet the urgent demand for increased production and a reduction in the unit cost of manufacturing moldings.

Altho the regulation, square-head molder with feeds ranging from 15-ft. to 40-ft. a minute was, and still is, a satisfactory machine for making short runs of odd and special moldings, millmen recognized long ago the need of a faster-feed machine for making standard patterns in large quantities. When fast-feed planers and matchers made their first appearance some years ago, and startled the lumber world by successfully producing high-grade dressed lumber, flooring, ceiling, etc., at feeds of 150-ft. to 300-ft. a minute, stock moldings were still being run at slow feeds and no better or more rapid means of manufacturing molding was offered to the trade until a few years later.

The problem of bringing out a fast-feed molder suitable for a variety of work naturally involved the overcoming of more serious obstacles than those encountered in perfecting fast-feed planers and matchers. The principles upon which the success of the fast-feed idea is based were the same in both cases, but the irregular shapes and deep cuts in molded work made it necessary to devise different types of multiple bit heads and cutters, different knife setting and truing devices, etc., than those employed on other kinds of fast-feed machines. All difficulties, however, were surmounted in time and, after passing thru the usual experi-

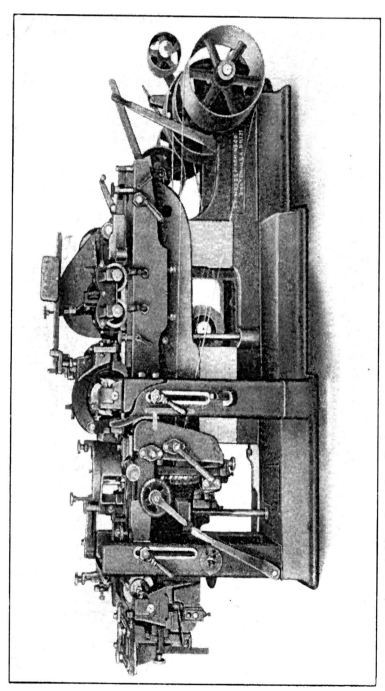

An outside, four-head, fast-feed molder fitted with round slip-on heads.

mental stage, the high-speed or rather fast-feed molder reached a degree of perfection equal to that of the present-day fast-feed matcher.

AN EXPLANATION OF MOLDER SPEEDS.

The underlying or basic principle which makes fast-feeds possible is to get more knives into action at each revolution of the cutterhead. Those who have given the subject more than casual attention know that when a square-head molder

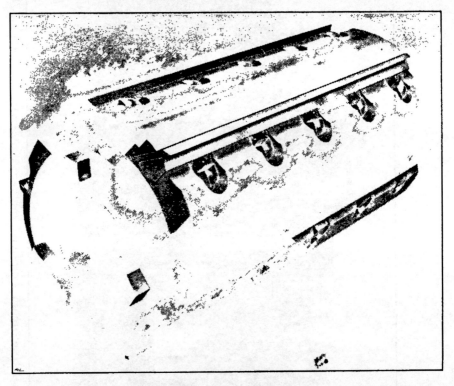

One type of six-knife, slip-on round head for top or bottom spindle of molder.

is set up with ordinary knives in the customary manner and put into operation, there is only one knife cutting at any one part of the stock. There may be a dozen knives positioned around the four sides of the head but no two cut alike unless by rare coincidence. This statement can be verified any time by observing the dust marks on the back edge of the knives after feeding a piece of material a few inches past the cutting heads. If only one knife of a kind strikes the material at each revolution of the head, there is only one knife cut per revolution; hence, a molder head rotating 3,600 r.p.m makes 3,600 knife cuts a minute.

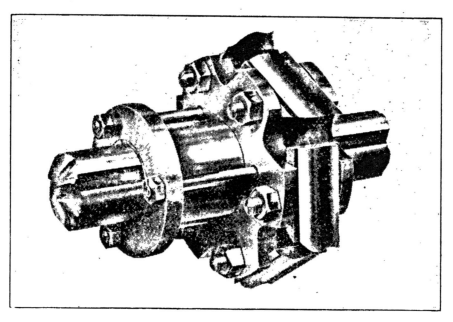

Self-centering "vise grip" type of profile head carrying high-speed steel, milled-to-pattern cutters.

Feeding stock at the rate of 25-ft. a minute (300-inches a minute) gives a ratio of 3,600:300 or 12 knife cuts per lineal inch which, when other conditions are right, results in a comparatively smooth finish. Suppose the feed is increased to 30-ft. a minute (360-inches a minute) it gives a ratio of 3,600:360 or 10 knife cuts to the inch, which means each knife will cut away a little more wood, resulting in slightly more chipping and deeper knife marks on the surface of the molding. The work may be passable but will not be as smooth as that run at the rate of 25-ft. a minute. The cutterhead speed might be increased to 4,000 r.p.m. which, with a feed of 30-ft. a minute, would give about 11 knife marks to the inch, but it is unwise to run molder cutterheads at such high speed because of possible vibration, heating of boxes, and the difficulty in obtaining a good running balance with the knives and bolts.

The principle of fast-feed machines is to get more knives in action per revolution rather than to increase the revolutions per minute of the cutterheads. Two knives cutting instead of one doubles the number of knife cuts per revolution, four knives increases the number fourfold, and six

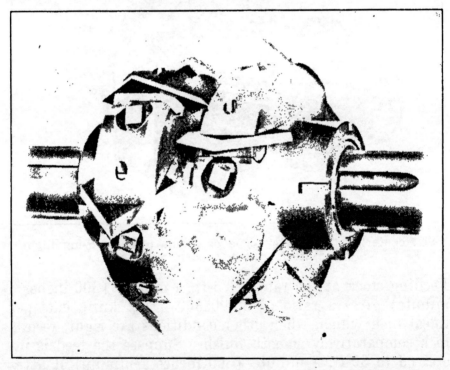

Two universal chamfer heads set side by side. These cutters may be adjusted to cut practically any bevel desired.

knives sixfold, etc. When the number of knives of a kind on a cutterhead are doubled or increased four or sixfold, and all of them are brought into equal action by means of knife-truing devices, described later, the rate of feed is

One type of fast-feed molder fitted with hopper-feed attachment.

increased in direct proportion without affecting the quality of work.

There are four factors, each bearing relationship to the other, which when taken together, determine the quality

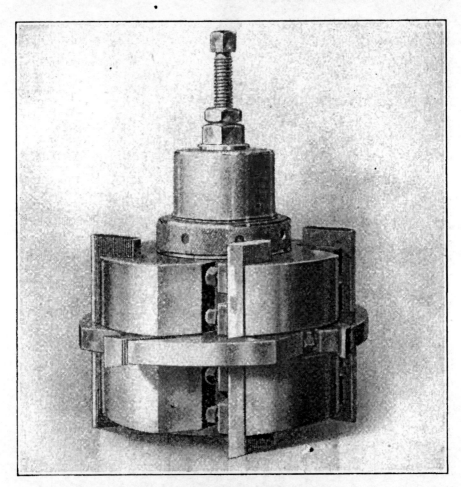

A three-disc, combination head for grooving heavy planks, etc. Notice enlarged section in middle.

and quantity of work that can be turned out on a molding or planing machine. They are the speed of the cutterhead in r.p.m.; the speed of the feed in feet or inches per minute; the number of knives in action on the head; the number of knife marks (whether visible or not) per inch on the finished molding. The relationship which these factors bear to each other is expressed by the following rule: The product of the r.p.m. of a cutterhead multiplied by the number of knives in action, divided by the rate of feed in inches per minute equals the number of knife marks per lineal inch on the finished work. Perhaps a more clear way to express this relationship is by equation form, as follows:

R=r.p.m. of head.
F=rate of feed in inches per minute.
K=number of knives on head.
X=number of knife cuts per lineal inch.

$$X=\frac{RK}{F} \qquad F=\frac{RK}{X} \qquad R=\frac{FX}{K} \qquad K=\frac{FX}{R}$$

Expressed in proportion this amounts to

$$RK=FX \quad \text{or} \quad \frac{F}{K}=\frac{R}{X}$$

Note: For the benefit of those not versed in algebra, the times sign (x) is omitted between letters which are to be multiplied, RK meaning R x K; FX meaning F x X, etc.

Now, by substituting known values for any three of the letters in the above equations, the third can be calculated by the simple formula given herewith, and one can determine all the factors upon which depends the quality of work, and the speed at which it is run. That is, exact calculations can be made as to the speed of cutterheads, speed of feed, number of knives in action, and the degree of finish (number of knife marks or cuts to the inch). This eliminates "cut and try" methods and the usual experimenting with different size pulleys and speeds which take up so much valuable time, spoil good lumber, and entail extra expense for labor and supplies.

When making calculations as to the possibilities of various molding and planing machines, as previously described, one must keep within certain limitations in regard to the figures used to represent the speed of heads, rates of feed, and the number of knife cuts per lineal inch of stock. No hard and fast rules can be laid down in this matter but the following recommendations are based upon results of ex-

Special combination head for working one of the many unique patterns run at the National Cash Register Company's plant.

periments and common practice in mills thruout the country: 3½-in. square heads—3,700 to 4,000 r.p.m.; 4¼-in. square heads—3,300 to 3,600 r.p.m.; 6-knife round heads—3,000 to 3,200 r.p.m.; 8-knife round heads —2,800 to 2,900 r.p.m.

Rates of feed, especially on fast-feed machines, vary according to the quality of finish desired, the kind of machine and its equipment, and the facilities available for getting material to and from the machine. A great deal of pine and fir flooring of good grade is made on fast-feed matchers

A five-head molder which can be used for either slow, heavy, or fast-feed work. In addition to the usual four heads, it has a top profile spindle which permits the first head to be used as a sizing or planer head.

equipped with automatic feeding tables, at speeds ranging up to 300-ft. a minute, but molding is seldom run at half this speed on fast-feed molders.   In fact, the average high-speed molder is generally rated to feed up to or near 100-ft. a minute which is about four times as fast as high-grade molding is made on the old-style, square-head machine.

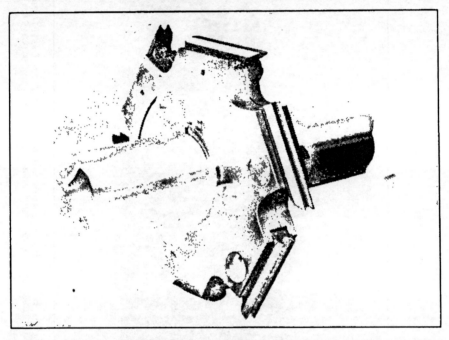

Profile beader head fitted with high-speed steel, formed knives.
Used to work beaded ceiling, etc.

Faster feeds are possible, but not always practicable, because they do not give the operator sufficient time to properly grade and turn the stock as he feeds it.

There is also a limit to the number of knife cuts permissible to the lineal inch of feed. Generaly speaking, the more knife cuts per inch, the smoother the work; but there is a recognized limit beyond which it is unprofitable to go. This limit is about 18 cuts to the inch. If there are more cuts than this the knives do not "bite" into the material deep enough to cut efficiently and the result is a rubbing, scraping action which rapidly dulls and heats the knives. Anything between 12 and 18 cuts to the inch produces a nice smooth surface; 10 to 11 cuts to the inch often gives a passable finish, but less than 10 to the inch invaribly shows the knife marks badly, in addition to chip-

A groove head for flooring. Notice the method of adjusting and locking the groover and its holder.

ping and tearing the grain wherever knots or cross-grain are encountered. This, in substance, is the possible and impossible, the practical and impractical in molder speeds and quality of finish. It reduces what formerly has been more or less an uncertainty and mystery to a simple matter of true facts and figures.

CUTTERHEADS AND GENERAL EQUIPMENT.

High-speed molders are built in both inside and outside models. The inside type is constructed with four, five, or six heads and usually fitted with permanent round top and bottom heads or cylinders, each carrying four or six knives, as desired. The top and bottom profile arbors at the rear end carry interchangeable heads for irregular work. Four-side slotted heads may be used on the arbors for special work, using ordinary cutters at a slower feed.

Heads carrying cutters for fast-feed work are of special design and possess the self-centering slip-on feature which permits fitting them up completely in the grinding room while the machine is in service on other work. The cutters are made of high-speed steel ground to shape in the usual manner, or milled on the back to the profile of the mold-

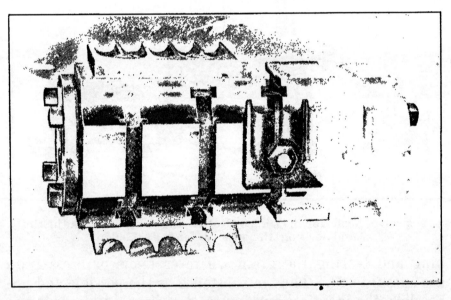

Fig. 60. Transverse T-slot head carrying formed knives for multiple work.

ing, as shown in Fig. 60. In either case, the cutting edges should be jointed while in motion to bring them all into exactly the same cutting circle. Each kind and type of cutter has its specific advantages. When there is a large quantity and wide range of work it is often profitable to have machines and equipment of different kinds and makes, each selected with a view of its adaptability for certain lines of work. There are machines which have a much wider range than others and consequently can be used for

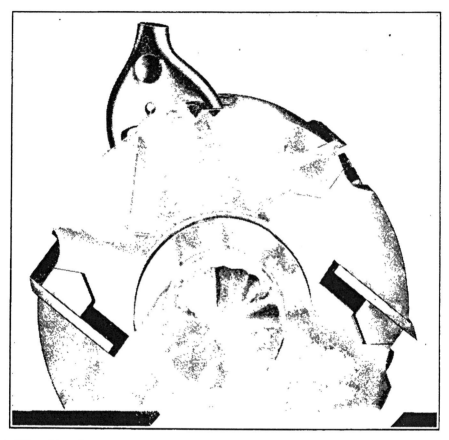

Fig. 61. One type of knife-setting jig for setting straight thin steel knives on round heads.

a greater variety of work, yet for some particular kinds, the machine with the lesser range is more satisfactory and efficient.

In selecting cutterhead equipment, it is advisable that every arbor for slip-on heads should be the same size, regardless of whether the machine is a molder or matcher. Top, bottom, side, and profile arbors should be exactly uniform size. All side-head sleeves and collars should receive all cutterheads and discs, irrespective of the type. This is easily accomplished as manufacturers will fit machines with any size arbor and furnish heads and discs

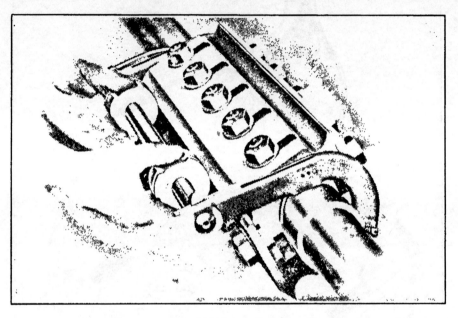

Fig. 62. Radial gage for setting knives on square, round, or three-wing heads.

likewise. The importance of this uniform equipment cannot be overestimated. It effects a great saving in set-up time, cost of cutters, and first cost of equipment.

To obtain best results and maximum number of productive hours from a fast-feed molder, adequate. equipment must also be provided for balancing, setting, grinding, and jointing the cutters. Cutters cannot be set on round heads or special high-speed heads with an ordinary molder rule. One type of device for accurately gaging the projection of straight knives in round heads appears in Fig. 61; another is shown in Fig. 62. A setting and balancing stand fitted with a templet for setting irregular molding cutters quickly and accurately on any type of cutterhead is shown in Fig. 63.

It is quite imperative that each one of a set of knives placed in a high-speed head be of the same make (therefore uniform temper and grade of steel) the same thickness, width and length, and the same bevel and weight. When setting and clamping up a set of knives in a round slotted head, the clamp blocks should never be set down hard

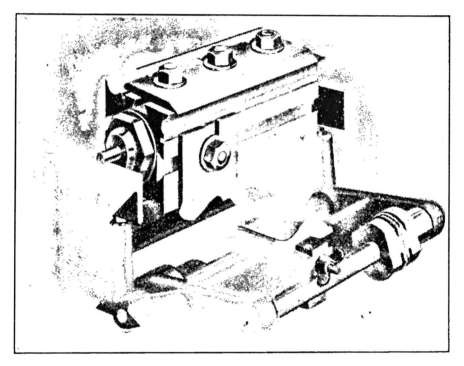

Fig. 63. A stand for setting and balancing irregular cutters on
any kind of cutterhead.

against any one knife until all of the knives are in the head and clamped down fairly tight. When the first knife is put in, lock it just tight enough to hold it in place until all the cutters are in the head, then each one should be adjusted and the clamp screws set evenly until the head has been gone over three or four times, before the clamping process is finished. If a cutter is clamped to the limit at once, while the clamp screws in remaining slots are slack, uneven strain is set up which sometimes causes poor work and hot bearings. Another method of setting and clamping knives in round heads, when making a change, is to

Fig. 64. Jointing straight thin steel knives on a round head while it is running at full speed.

loosen and remove only one knife at a time and immediately replace the removed knife with a sharp one and tighten to the limit before another is ever loosened. The latter method is a little quicker than the former but not always practicable.

JOINTING CUTTERS ON HIGH-SPEED HEADS.

After a set of high-speed knives are set as accurately as possible to get them with the aid of modern gages and knife-setting devices, jointing at full speed is the next operation. Since jointing is the final operation before the

Fig. 65. Jointing thick knives on square head.

head and knives are placed in service, it is a really important one and must be performed with great care and skill. There are special stands or arbors on the market for jointing the knives of slip-on heads, and altho many are in use, opinion is divided among practical millmen as to whether a cutterhead jointed on an arbor in the grinding room and then transfered to the machine will produce as nice a finish as a cutterhead jointed right on the machine spindles.

When the knives on cutterheads are jointed on a jointing stand, the jointing arbor, the machine spindles, and the
self-centering sleeves in the heads must all be mechanically

Grinding straight knives with portable grinder which is
moved back and forward on a horizontal dove-tail slide or bar
secured rigidly on the yoke of the machine in accurate alignment with the cylinder.

perfect and in tip-top condition; otherwise, the jointing
will never come right. The jointing arbor must be the
same diameter as the machine spindles and must run
smoothly and quietly in well lubricated, massive boxes at a
speed corresponding to that of the molder heads. If either

the jointing arbor or any of the machine spindles are the least bit out of true or balance, or if either has at any time been trued-up after having been in service, the jointing will not come right when a head is transfered from one to

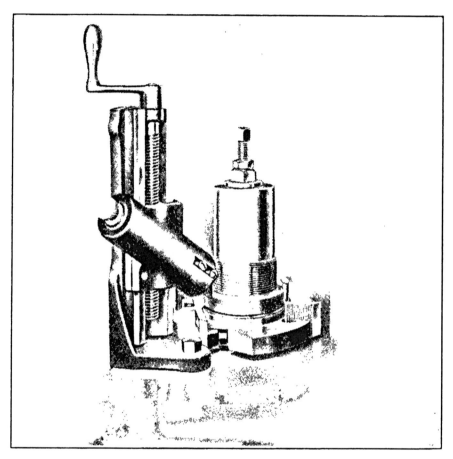

Fig. 66. One type of side-head truing device in position for jointing a 4-knife side head for square-edging thin stock.

another. When all of these little things are taken into account, anyone of which will destroy a perfect jointing, it is easy to understand why many practical millmen prefer to joint cutters while the cutterheads are positioned and clamped on the machine spindles. A good jointing arbor, however, is of inestimatable value in every grinding room whether used for jointing cutterheads or not. It serves as an excellent machine for testing the running balance of all slip-on heads before they are put in service on the machine spindles. By trying out the heads for running balance in the grinding room, much valuable time is saved because any error in balance can be detected and corrected before the head is put on the machine. Without this preliminary running test, errors in balance are often not dis-

Fig. 67. Showing formed stone and holder for jointing irregular shaped molder knives on top or bottom head.

covered, if at all, until after a head is placed in service and then it is generally a temptation to let it go. A poorly balanced head on a molder spindle results in poor work, hot boxes, and eventually worn bearings.

Best results in jointing are obtained by doing the work

while the machine is warmed up and the journals are flooded with oil. Joint lightly always. Let the "touch" of the stone, the faint whir of the knives and the appearance of a small, dark-red spot on the end of the stone tell when

Fig. 68. Another type of side-head jointer and a special four-wing, fast-feed head fitted with self-centering sleeve and thick high-speed steel cutters.

the stone is in action rather than crowd the jointing until a stream of red sparks fly into the air. Heavy jointing is ruinous because it produces such a heavy heel that the knives pound and raise the grain instead of cutting freely as they should. Also, the edge of the cutters will burn if the feed stops for a moment. On the other hand, when the knives are jointed lightly, they will stand several jointings (each of which renews the cutting edge) before they require regrinding.

The device generally used for jointing straight knives on top and bottom heads at the machine consists of a per-

Fig. 69.   Side-head jointing attachment in position for jointing the formed cutters of a matcher head.

fectly straight slide bar with a movable carriage accurately fitted thereto which carries a jointing stone.   The slide bar is set absolutely parallel to the cutting cylinder and secured rigidly in place so the jar or vibration of the machine

Fig. 70. Showing one type of jointing device attached to slide bar on a jointing and setting stand. Self-centering heads are set up, tried for running balance and jointed at full speed on the stand, and then transferred to molder spindles.

cannot effect it. During the process of jointing the carriage is moved slowly from end to end while the cutterhead is revolving at full speed. Provision is made for moving the jointing stone toward or away from the cutterhead and for holding it firmly to the work, see Figs. 64 and 65.

The devices for jointing straight knives on side heads are similar to those used for top and bottom heads. They are designed to be fitted to the machine frame at a point near each side head. The jointer-stone carrier works snugly on a vertical slide which sets parallel to the side head. Both vertical and horizontal adjustments are provided so the stone can be moved to-and-from end to end of the slide with ease and accuracy, see Figs. 66, 68 and 69.

There are two methods commonly used for jointing irregular shaped knives. One is to prepare a formed stone, one edge of which must be made exactly the same shape as the profile of the molding. This formed stone is

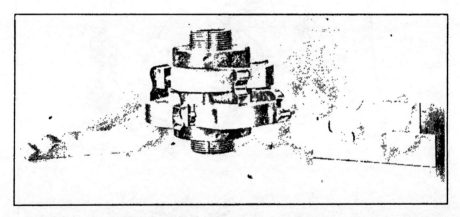

A two-disc combination on self-centering clamp sleeve. Steel jointing form, in two sections, appears at right.

clamped in the stone carrier and positioned so it lines up exactly with the molding cutters on the head, see Figs. 67 and 68. The carrier is then clamped fast to the side bar, the stone backed away from the knives, and when the head is running full speed the stone is advanced slowly and carefully until it comes in contact with the edges of the whirling knives. If the knives have been ground and set with extreme care and accuracy a slight touch with the

One type of pedestal head grinder for grinding knives on self-
centering side and profile heads.

jointing stone will be sufficient to bring them all into a uniform cutting circle and make each do an equal share of cutting. The other method consists of using a device like that attached to the jointing.stand shown in Fig. 70. A steel templet or pattern is clamped to the bar and a pointer which is arranged to travel on this formed pattern guides a narrow jointing stone in such manner that the exact profile of the pattern is reproduced on the cutters. The steel patterns or templets can be purchased from manufacturers or made in the grinding room to suit the form of cut desired.

After the heads have been fitted and positioned on their arbors, the operating of a fast-feed molder does not differ greatly from that of a slow-feed machine. The adjusting of the rolls, bed, guides, and pressure bars is done in practically the same manner as on the old-type machine. However, in the case of cutterhead bearings there is a difference because, when using high-speed heads and jointed cutters, it is necessary to always have the bearings absolutely tight and well lubricated in order to keep all knives cutting in such a manner that they will produce a perfect surface on the finished molding. The adjustment of the cutterhead bearings, however, does not have to be made as often on a fast-feed machine as on the old-style, slow-feed machine because the journals are larger, caps are more securely clamped, and the massive boxes are mounted in heavy, powerful yokes which positively hold them in place when once adjusted.

In making short runs of special molding the method used is virtually the same as on the square-head machine. Square heads are slipped onto the spindles and clamped by a self-centering device. The regular knives used in detail molder work are used in the same way as on the old-type machines and about the same rate of feed is carried.

The fast-feed molder is not confined to molding work alone. Owing to its rigid design and powerful feed works,

A six-head, fast-feed, inside molder. Bottom and top surfacing heads at front, top and bottom molding heads at rear.

it will do the ordinary work of both a fast-feed matcher and double surfacer and is frequently used as such. It is also particularly adapted for heavy work when fitted with square slip-on heads on account of its heavy, powerful construction. Since its greatest advantage, however, lies in making long runs of stock molding, a good supply of slip-on heads and high-speed steel cutters should be kept on hand at all times to make the various patterns of molding regularly manufactured.

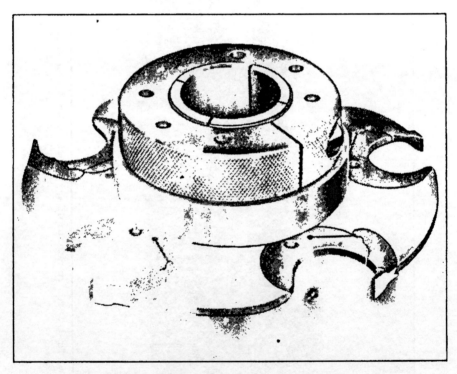

Inserted, swaged-tooth ripping saw, with clamp collar and self-centering sleeve, for use on a molder spindle.

CHAPTER XIII.

KNIFE MAKING.

The designing and making of knives is an art that should be thoroly understood and mastered by every mechanic who aspires to become a first-class molderman. Heretofore, the opportunities for an apprentice or the uninitiated to learn knife making have been few indeed. More or less secrecy has always been thrown around the principles and correct practice of designing, shaping, and tempering molder knives, with the result that many molder operators are not acquainted with the most up-to-date methods in this interesting and important part of their trade.

Previous to the advent of modern high-speed steels, which require no heat treatment, carbon steel was used exclusively for all straight and irregular molder knives. Because of the comparative low cost, and its satisfactory performance on detail and short-order work, carbon steel is still widely used in woodworking factories. It is usually purchased in slotted blanks already cut to length and width, and beveled at the cutting edge. It also comes in the form of rectangular bar steel of various sizes. As a rule, steel for slotted knives is bought in slotted blanks about ⅜-in. thick, while stock for spike knives comes in straight bar steel ¼-in. to 5/16-in. thick and ¾-in. to 1½-in. wide.

When ordering knife steel for one or more molders in a plant, always specify the same thickness, especialy in the case of bar steel for spike knives. It is best to stick to one brand of steel as long as that particular kind is giving good results. Cheap steel is expensive at any price and should never be considered under any circumstances.

The first thing in knife making is to determine the

design and size of cutter or combination of cutters most suitable for the work at hand. To design a knife intelligently, one should know how the molding is to be run and the kind of wood that is to be worked. The design of a single knife or group of knives for any particular pattern depends more or less on the number of cutterheads to be employed and whether the molding is to be run face-up or face-down, flat, on edge, or at an angle, etc. If the material to be worked is curly-grained hardwood, or if it must be molded across side or end grain, the knives will, of course, have to be designed to meet such conditions. A number of common and special knives for various kinds of work have already been illustrated and described in preceding chapters. The reader is further reminded of what was said in Chapter III about the advantages derived by using sectional knives instead of a single solid knife, especially when making complicated patterns. When planning a new knife, avoid all inside corners and other shapes that require the use of a file for sharpening the cutting edge. A knife that must be filed requires a filing temper, therefore it cannot be depended on to hold a sharp cutting edge. It dulls rapidly and often loses its true shape after being sharpened a few times. Knives that can be completely ground to shape on a wheel are more satisfactory in every way. They are easier to sharpen and can be given a harder temper along the cutting edge.

Knives to cut wide sweeping curves such as shallow ogees, ovals, etc., are generally made in one piece rather than in sections because, in cases of this kind, one knife is easier to make and set than two. Under such circumstances, a single solid knife is all right provided the cut is not too deep and the material is comparatively straight grained. Otherwise the cut should be divided between two or more knives positioned on different sides of the head.

Moldings which are rabbetted to overlap base, wainscoting, or any woodwork ¾-in. or more thick, are almost in-

variably made as shown in Fig. 71 in preference to the manner illustrated in Fig. 72, because the former method saves
considerable lumber. When molding is turned at any angle,
and run in this manner, the knives must be designed accordingly. Sometimes by turning a molding, as shown in Fig.
71, a light under-cut is necessary, whereas, if it is made
like Fig. 72, no under-cutting is required. This is another
case where one must use sound judgment in deciding upon
the best practice rather than follow any hard and fast rules.
Usually if there is a large quantity of molding to make, the

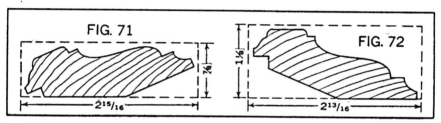

Figs. 71 and 72. Notice the saving in stock effected when molding is run as shown in Fig. 71.

saving in lumber is of far greater importance than the
little extra trouble required to tilt the side head to make a
small under-cut.

Never attempt to make a deep cut with a long, slender
knife. Thick steel knives should be used for making deep
cuts, and wherever possible, each knife should be of extra
width to give added strength. Before making a new knife
be sure the steel blank is neither too long nor too short for
the head. If too long it can be cut down, of course, but if
too short it cannot be used. With the design and size de-
termined, the next thing in order is to grind the steel to
correct shape and bevel to produce a knife which will cut
the desired profile. Knife shapes for bevels and straight
cuts are simple enough, but a knife which will cut a true
quarter or half-round, a cove, ogee, reverse ogee, or any irregular curve, is not so easily ground to shape until one
has had some practice at the work.

Owing to the angle at which knives on square heads are presented to the work, they must project farther from the lip of the head than the straight-down measurement of the corresponding cuts which they make, see Figs. 73 and 74.

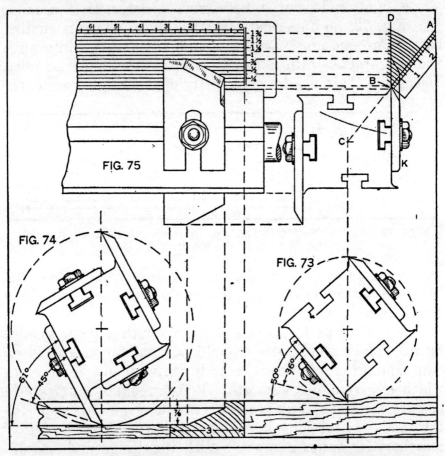

Figs. 73 and 74. Notice how the cutting angle changes with depth of cut. Fig. 75. Method of laying out molder scale.

The angle at which molding knives do their cutting is by no means constant, but varies slightly according to size of the head, and considerably according to the knife projection or depth of cut. Thus, in Fig. 73, a surfacing knife on a 4-in. head (reduced to scale) swings past the point of its deepest cutting at an angle of about 50 degrees, while in Fig. 74, a knife cutting ⅞-in. deeper on the same size head finishes at an angle of 61 degrees.

The exact amount of knife projection required for cutting different depths is obtained by making a full-sized

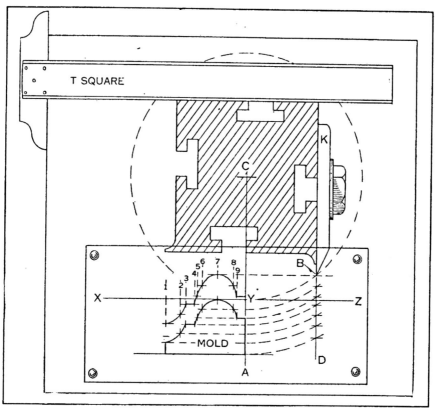

Fig. 76. Method of laying out knife profile with small drafting outfit.

layout of the cutterhead with at least one knife in position, see Fig. 75. Make an allowance of about 3/32-in., or enough to properly clear the knife bolts, for the projection of surfacing knives, as at B, Fig. 75. From the center of the head draw line C B A, and on line B A start at B and lay off ⅛-in. divisions from a rule. Continue the knife line K B to D. Then with the compass point at C, extend the ⅛-in. divisions from line B A to the knife projection line B D. The divisions on B D represent the true molder scale for this size head. Close examination of this layout discloses the fact that while all of the divisions on line B D are slightly more than ⅛-in. in length, they gradually decrease in length as they recede farther and farther from the head. This point is mentioned to explain clearly why no molder scale can be reversed.

Now, when the molder scale is laid out, as just described, it can be transferred to a hardwood, metal, or celluloid gage as shown in Fig. 75, or it may be scribed on either a little brass T-square or the edge of an ordinary rule. It can then be used for laying out new knife shapes to guide one in both the grinding and setting operations, see Chapter IV., "Setting Up a Molder".

Fig. 76 shows a method of laying out knife shapes with a small drafting outfit. A full-size head is laid out permanently on a small drawing board. The sides of the head are square with the edges of the board, and knife line K B D is parallel to the line of actual measure C Y A. The horizontal base line X Y Z intersects vertical line C Y B at right angles and is tangent to the cutting circle of the surfacing knives. A tracing of the full-sized molding is tacked to the board so that its inside edge lies on line Y A and its highest point touches line X Y, as shown in Fig. 76. Now, with the T-square, draw horizontal lines thru important parts of the molding and let them intersect vertical line Y A. With the compass centered at C, extend these lines to the knife projection line B D. At points on the molding-outline where the horizontal lines intersect, square up as at 1, 2, 3, 4, 5, etc. Then square back across with T-square from the intersections on line B D and, where the vertical and horizontal lines intersect directly above the molding, the correct knife profile can be traced, as shown in Fig. 76. This method of laying out knives is accurate but slow, and therefore only recommended for study practice and use in drafting rooms.

Expert knife makers use neither the drafting system nor the sticker rule for obtaining correct shapes, but immediately begin grinding the knife to shape on a coarse, free-cutting wheel without any preliminaries whatever. First the knife is ground to a shape exactly the reverse of the molding, then ground deeper to make allowance for the greater knife projection. At this stage of the process one is guided by good judgment and a practiced eye. As the grinding nears completion the knife is taken from the wheel repeatedly and held over the drawing, or fitted to the sample at the angle which it cuts. The angle is changed for deep and shallow cuts as dictated by keen judgment. By sighting down over the edge of the knife to the outlines of the drawing or sample, one can tell when the shape is right, and when right, the clearance bevel is ground. This method of grinding knives to shape is not guesswork, as one might presume, but is a matter of skill and practice, being used by some of the most accurate and fastest moldermen in the country. By using this system, a rapid workman will have a knife half completed in the time that it ordinarily takes to lay out a knife shape with pencil and paper. .

Before leaving the subject of knife shapes, it might be well to explain a little point over which there has been some argument and speculation among moldermen. The question is, why a straight miter or bevel cannot be cut with a straight-edged knife. The fact of the matter is that a straight-edged miter knife cuts a slightly convex or curved miter, while to produce a really straight miter or bevel the knife must be slightly curved (convex) instead of straight. This apparent paradox is due to two things: the elongation of the cutter to conform to the elongated molder scale, and the fact that the cutting angle of the knife changes according to the depth of cut. The layout of a miter knife in Fig. 77 shows the slight curvature clearly.

Another thing worth remembering about knife' shapes is that a knife ground to cut a perfect quarter-round or half-round is part of a true elipse. One method of laying out knives to cut perfect quarter and half-rounds appears in Fig. 78. A wood-turning of the proper diameter, however, is probably the best thing to use in trying out a quarter or half-round knife to prove its accuracy before putting it in service. Moldermen who have occasion to make a great deal of round molding will find it advantageous to prepare a stick with different diameter sections turned along its length for this purpose.

Knives for round heads can be drafted and ground to shape in a manner similar to that just described for square heads. The principle is exactly the same in both cases. The proper bevel for the edge of molding knives depends largely upon the size of the head and the angle at which the knives cut. There must always be more than enough bevel for back clearance in order to give an acute cutting edge and prevent any possibility of shavings catching between the heel of the knife and the finished molding. Knives for comparatively shallow cuts should be beveled on the edge about 33 to 36 degrees, while those for deeper cuts may be given an edge-bevel of 40 to 45 degrees, see Figs. 73 and 74. The reason for the shorter bevel on long knives is to give greater strength at the cutting edge. The amount of bevel for side clearance need never be more than 5/16-in. or 3/8-in. to the inch, or about 1/8-in. to the thickness of an ordinary knife.

Knives made of carbon steel must be given heat treatment, followed by quenching in oil, water, or some liquid solution to give them the proper degree of hardness or temper along the cutting edge. A knife that has slender points should not be ground to finish size until after receiving heat treatment. A new knife edge should be left thicker at all narrow points and corners so that when it is being heated at the forge these points will not be so likely to heat too rapidly and burn before the rest of the knife becomes red hot. If the knife is to be heated in an open forge, use old coals or charcoal for the fire in preference to fresh coal because the latter may contain sulphur or other

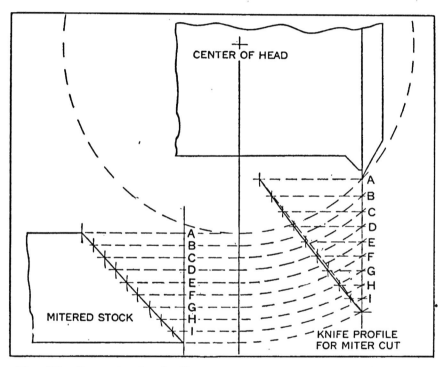

Fig. 77. Layout of a knife to cut a miter. Notice the slight curvature of knife edge.

chemical properties that may injure the steel. Before putting a knife into a forge fire, it is important to have a heaping bed of red hot coals. Lay the knife face up on the hot coals and concentrate the greatest heat a little back of the beveled heel. Never heat the cutting edge first. See that the heat plays uniformly along a line about ½-in. or so back of the heel. This uniform heating, which is so important, is accomplished by turning and shifting the knife with the tongs. Operate the bellows slowly to give a quiet but positive draft until an even dark red appears along the entire edge of the knife and ¾-in. or more back to the edge. Heating along back of the heel, as directed, permits the heat to spread more evenly and prevents the sharp cutting edge from overheating. If perchance a slender point on the edge heats too rapidly, withdraw the knife and cool that point either on cold iron, in tallow, or oil, then replace the knife and continue heating it. Never let a knife lay on or in hot coals to absorb heat, but try to keep its temperature rising and quench it at a rising heat.

As the knife gradually becomes a little brighter than dull red, push the entire edge into the coals, and then when it is a cherry red (a shade between dull and bright red) withdraw the knife quickly and plunge it point downward into a bucket of linseed oil, fish oil, or some suitable quenching liquid, and stir it vigorously to cool quickly and thoroly. When the knife is cool, wipe it dry and try the beveled cutting edge with a file. If it files easily, it is too soft. Should the file glaze over it like glass and not bite, the edge is then too hard and must be tempered to reduce the degree of hardness. When the file takes hold on the edge and bites with difficulty, the edge is just right. A knife that is found to be too soft after heating and quenching, as described above, has evidently not been heated hot enough or else not quenched quickly and thoroly while at a rising heat. A knife that is a little too hard can be easily tempered. First brighten the edge and face with sandpaper or some abrasive; heat it some distance back from the edge on a red hot iron or over the hot coals in the forge. After a little heating, colors will begin to appear. When a yellow or straw color spreads to the cutting edge, remove the knife and cool it. This usually gives a good temper for wood cutting and the edge can barely be scratched with a file.

There are several different ways of quenching steel after

it has been heated to a cherry red. Often the method is varied somewhat to suit the kind of steel or quenching liquid available. Some knife makers prefer about ½-in. of oil on top of the water. They dip the knife into the oil slowly, heel first, passing it slowly thru the oil and into the water until the knife becomes black. Then it is quickly withdrawn and polished on the edge and face. Usually enough heat remains in the part held by the tongs to temper the edge. The colors begin to appear as the knife is held in the air, and when a yellow spreads to the edge, the knife is cooled in water. This is a rapid method because only one heat is used for both the hardening and tempering.

In the absence of oil, water is sometimes used for quenching but is not recommended because it may crack the steel or make it so hard and brittle that it will crack when in service. If water must be used, the knife should be slowly dipped into it, heel first, but only at the surface however,

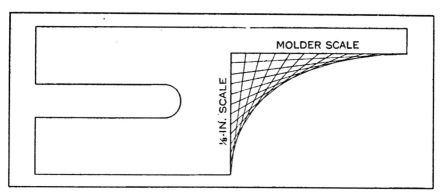

Fig. 78. One way to lay out quarter and half-round knives.

then withdrawn and dipped again, each time a little deeper, repeating the process until the knife is black. Then polish quickly and temper as described.

There are a number of so-called secret solutions for quenching steel, and while some give good results, they are all more or less expensive and very few excel good linseed oil or a combination of fish oil, linseed oil, and tallow. Plain lubricating oil can be used with fairly satisfactory results. Oil of one kind or another is prefered to water on account of its milder action as a cooling agent.

After a knife has been properly tempered the edge should be ground smooth and sharp, and whetted lightly with a whetstone. It is then ready for use. To preserve the exact shape of irregular stock knives, patterns to fit the cutting edge can be made of sheet metal and filed away with either the knife, the molding sample, or set-up templets.

Bar steel for spike knives is generally cut into lengths suitable for knives by grinding deep grooves across the side with a thin emery wheel and then breaking off the pieces in the vise. When following this method, grind the grooves on one side only, as shown at A, Fig. 79, not on both sides as at B, Big. 79. The latter is a very wasteful practice and requires considerable more grinding to attain the cutting bevel.

The practice of heating and swaging or spreading the cutting end of spikes by hammering, see Fig. 80, before they are ground or tempered is very good because it gives a wider cutting edge and leaves less to do when beveling the edge. It is also claimed that hammering hot steel during the swaging process tends to compress it and make the cutting edge more tough.

Knives for cutting wood across the grain are often twisted or forged to a curve to give the cutting edge a shear cut similar to the angle of knives on automatic lathes and tenoner heads, see Fig. 81. A shear cutting knife with a hard edge produces much smoother work than straight knives. Twisted knives are sometimes used for making deep, perpendicular or nearly perpendicular edge cuts with the top or bottom heads. The twist gives the knife an acute cutting angle at the side, therefore it cuts easier and stays sharp longer than if it was flat and made a side scraping cut. Knives should always be designed to cut freely, especially at the sides, because if they-scrape or rub the wood, the friction causes them to heat and dull rapidly. A knife will never burn black if it is cutting freely, but it will do so in a few minutes if allowed to rub on the edge of stock or between two strips of multiple work.

Molder knives should always be arranged orderly in

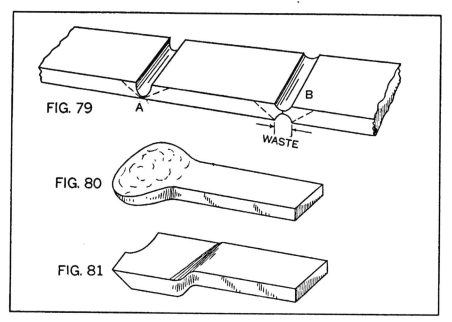

Fig. 79. Correct and incorrect way to cut off bar steel with emery wheel. Fig. 80. A spike knife, spread or swaged at end. Fig. 81. A twisted knife.

clean racks so that no time need be lost in finding particular kinds. In large up-to-date factories, knife racks, grinding wheels, forge, vise, balancing scales, set-up equipment, etc., are kept in a separate, well-lighted room called the grinding room. The knife making, grinding, balancing, and selecting of cutters for several molders is all done by an expert molderman. The machine operators simply do the setting up and tend to their machines. In small plants, however, the knife rack is often purposely placed near the molder so that while the operator is feeding the stock on one job of molding, he can also be picking out and balancing cutters for the next job. In some California factories, a knife rack is built right over the countershaft of each molder, and in front of the rack there is a narrow work table or shelf with drawers for bolts, waste, and other paraphanalia. The balance scales sit on the shelf ready for instant use. While feeding the molder on one order, the operator selects his cutters and balances them for the next job. This is one of the little conveniences that enable the "speed kings" to make so many set-ups a day.

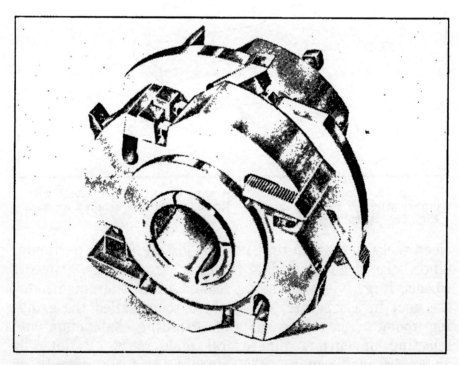

A three-disc side head tipped to show self-centering sleeve.

CHAPTER XIV.

BABBITTING HIGH-SPEED BEARINGS.

Babbitting the boxes, which carry high-speed spindles, is very particular work and calls for the exercise of both skill and good judgment. On some machines there is a brass name-plate bearing the words, "Never babbitt on a cutterhead journal; you may spring it, and once sprung it cannot be permanently repaired; use a babbitting mandrel." This is good advice and should always be followed when possible. The greatest danger in springing a journal is when the hot metal is poured directly onto it, and only on one side, to cast a half box. The sudden expansion during the time of pouring, and slow contraction later, is likely to produce a permanent set or slight bend in the journal which will cause trouble by heating and running badly.

When a journal, however, is wrapped with two thicknesses of thin paper and both top and bottom boxes are poured at the same time, there is scarcely any danger of springing it. Still there is an element of risk in the latter, hence the safest course is to always use a babbitting mandrel made of an old spindle, shaft, or stick of hardwood turned to the same diameter as the journal.

A mandrel should be wrapped with two thicknesses of thin paper and the ends pasted down with mucilage or photo paste. The paper enlarges the mandrel just enough to take care of the shrinkage of babbitt in cooling. Babbitt also casts more smoothly around paper than it does around a naked metal shaft. The mandrel must be carefully and firmly secured in exactly the same position that the cutterhead spindle assumes while running. In other words, if one or both boxes for a top, bottom, or profile spindle are to be babbitted, the mandrel is placed level and square with the machine bed. If only one box is being cast, it must line up with its mate. When both are to be cast, the mandrel should always be placed as nearly in the center of the boxes as possible. In adjusting a mandrel for side-head boxes, set it plumb with the machine bed.

It is generally much more convenient, when preparing to babbitt side-head boxes, to detach the yoke in which the boxes are mounted and take it to the repair room where the work can be done in good light near the forge. Align the mandrel parallel with the planed ways of the yokes.

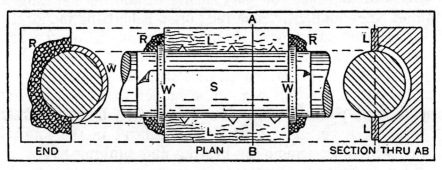

Fig. 82. Showing lower half of the box prepared for babbitting. S, babbitting mandrel wrapped with paper. L, L, liners. W, W, washers at ends. R, R, putty.

Clean out all old babbitt, especially that in the anchor holes and plug up any oil holes or chambers which may be in the bottom box. Oil holes can be easily plugged with bits of wood whittled round to fit them. Oil chambers, however, should be packed with waste and the opening covered with a piece of belting thick enough to fit in snugly between the box and the babbitting mandrel. This piece of belting backed with waste, not only keeps the hot metal from running into the oil chambers or between the mouth of the chamber and the mandrel, but also serves as a shim which assists in centering and supporting the mandrel in the box.

A very simple and easy method of centering and supporting a babbitting mandrel is to place a narrow strip of belting crosswise of the box, letting it extend less than half way around the mandrel to leave plenty of room for melted babbitt to flow in all parts of the box. After the box has been cast, these strips of leather may either be left in position or replaced with felt. When both top and bottom boxes require new babbitt lining, considerable time can be saved by arranging to pour both boxes at the same time. As in all cases, the edges of the boxes must be separated by liners to provide enough take-up adjustment for wear. A plan or top view of the bottom half of a bearing ready to be poured in the manner recommended appears in Fig. 82. The liners L, L are made of cardboard or thin wood and have small V-notches, as shown, to permit the melted babbitt to flow from the top half of the box down to the lower half. When the metal cools, it will be joined together at the V-notches, but these slender connections are easily broken apart by a few taps of the hammer on the end of the top box, or the prying action of a chisel. If only the top half of the box is to be cast, the liners are not notched at all.

After the oil holes and chambers are plugged and the babbitting mandrel is properly centered and clamped in place between liners, see Fig. 82, the cap (top half of box) is set over the mandrel and clamped or bolted down against the liners. Leather or cardboard washers W, W, Fig. 82, are then placed at each end and the end openings sealed with putty, clay, dough, or a mixture of asbestos and oil, see R, R, Fig. 82. Two open holes should be left in the top of the cap, one for pouring the metal and the other for a vent to permit the escape of entrapped air. Care must be taken that no moisture is present in the boxes when the metal is poured or a "blow-out" will result from sudden formation of steam. If the work is done in a cold atmosphere, the boxes and mandrels should be warmed before the metal is poured. This avoids chilling the hot metal and permits the box to shrink somewhat with the metal when both cool, thereby preserving a firmer connection between the two.

When everything is in readiness, the babbitt is melted in a regular ladle. In the absence of a proper instrument for taking the temperature, one must simply use good sense in heating the metal, not too hot, but hot enough that it will run freely into all parts of the box before it begins

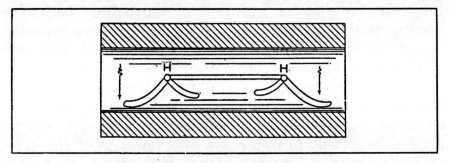

Fig. 83. Oil channels spread in direction of rotation from supply holes, H, H.

cooling. The temperature at which babbitt should be poured is about 450 to 460 degrees C. Pour the metal in a steady stream, being careful not to let any of the top slag enter the box and be sure to have enough metal to fill the entire box with one pouring. Do not disturb the newly cast box until it has cooled; then take the boxes apart and smooth off all rough or sharp edges with a chisel or rasp, and babbitt scraper. Open the oil holes, and chambers if any, and cut oil channels from them to points near the ends of the box, as shown in Fig. 83. The function of the oil channels is to facilitate the flow of oil to all parts of the bearing, therefore, they should lead outward from the supply holes in the direction of rotation. Their edges and those of the boxes should be rounded because sharp edges tend to scrape oil off the journal and prevent its proper distribution.

Every babbitted bearing must be carefully scraped to fit

its journal perfectly. Otherwise, the journal will bear only on high spots and all lubricant will be forced into the low places where it will not do any good. Under such conditions, overheating is certain to occur, even tho the bearing is flooded with oil. In order to know where to scrape, give the journal a very thin coating of red lead paint and turn it around in the boxes. The bright spots show where the surface must be scraped. When the fit between journal and boxes appears to be perfect, bolt the latter in place, put on the belts, and run the spindle up to speed for a few minutes. Take the boxes apart again, and the bright places caused by friction show exactly where to do the light, final scraping. This all takes time, but is worth it, because it p r o d u c e s smooth-running, non-troublesome bearings which are so essential to good molder work.

When adjusting new boxes to a high-speed journal, do not make the mistake of clamping them down too tight. Leave a little play at first and run the spindle at full speed for a while to let it warm up. The expansion from

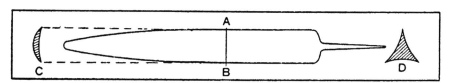

Babbitt scrapers are usually made by grinding a half-round file as shown at C, or a three-cornered file as illustrated by cross-section D.

heating may be enough to take up all lost motion, but if it is not, then tighten the boxes just a trifle while they are warm.

In babbitting a box which receives a grooved journal (grooved to prevent side play in the spindle) the usual practice is to give the journal a good coat of white lead paint, and then use it as a mandrel for molding the babbitt lining. A wood mandrel with corresponding grooves can be turned by an expert turner if a templet is provided, to fit accurately into the grooves. The wood mandrel is recommended because, if a grooved journal is sprung the least bit, it is practically ruined forever.

The lubrication of high-speed bearings is of great importance. To run properly, a journal and its bearings must always be separated by a thin film of oil. It is quite imperative that good oil be used and the supply be kept constant in order to maintain perfect lubrication. It is often necessary, especially on large bearings, to introduce oil from the bottom as well as the top. If there is no provision for an underfeed oil system, and one is desired, drill a small hole through the bottom of the box and tap it for a small U-shaped pipe. The short leg of the U should then be screwed into the bottom casting and the long leg fitted with a sight feed or plain metal oil cup. Oil channels must then be cut in the bottom box the same as in the cap.

CHAPTER XV.

Leather belting of good quality is undoubtedly the best for driving molder cutterheads. Light (22 or 24-oz.) two-ply belting is generally the correct selection for heavy-duty top and bottom head drives, but single-ply center stock is the logical choice for side head and all other drives on medium and small size machines. Next to leather is the four and five-ply woven canvas, impregnated with rubber or a similar substance.

To drive a cutterhead effectively, the belt must be very flexible so it will wrap around the small pulleys to good advantage. It must have great strength and be able to stand up under continuous high-speed service. When tighteners are brought to bear on belts to keep them at uniform tension, as they gradually become lengthened from continued service, the belts should be made endless. Otherwise, the ends are usually joined together with a tough wire lacing either hand or machine sewed. When cutting a new belt to net length, it is usually safe to allow about 1/10-in. to $\frac{1}{8}$-in. to the lineal foot for stretch. In other words, cut it just that much shorter than the actual tape measurement around the pulleys. Cut the ends of the belt perfectly square with a try-square. For ordinary wire lacing, to be put in by hand, use a punch which cuts a hole slightly less than $\frac{1}{8}$-in. Punch a single row of holes 5/16-in. from the end and let the holes be about $\frac{3}{8}$-in. apart on center. Make an even number of holes in each end, if possible, so the belt can be sewed in such a manner as to equalize the side pull of cross-over strands at each side of the center, see Fig. 84.

The method of sewing, shown in Fig. 84, is simple but effective and keeps the edges of the ends even. Always turn the grain or smooth side of a single-ply belt to the pulley and lace so that no cross-over strands show on the pulley side. Cut the wire about seven times longer than the width of the belt and start the two ends from the back side of the belt throught holes 1 and 1-A, respectively. Pull

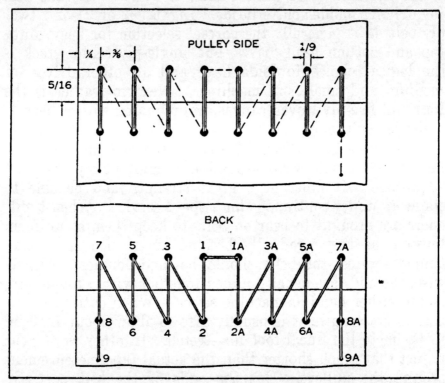

Fig. 84. Method of sewing wire lace by hand.

the ends evenly and draw them across the joint on the pulley side and thru holes 2 and 2-A, respectively, then thru 1 and 1-A again, then cross the pulley side thru 2 and 2-A. One strand is then crossed over from 2 to 3 on the back and the other over to 3-A from 2-A. Continue until the edges of the belt are reached, and then cross the wire over to an adjacent hole or draw it thru an extra hole, as shown at 9, Fig. 84. After lacing, hammer the wire down flat. The belt is then ready for service. Put single belts on so the point of the lap on the inside runs toward the pulleys, because the lap on the outside of a belt is most likely to come apart when the point is run against atmospheric pressure. Double belts should be put on so the points of the laps will run with the pulleys as both sides point in the same direction, see Fig. 85.

Belts should be kept clean and free from oil and grease. Mineral oils, in particular, rot leather rapidly. Where belting is liable to become oil soaked, mechanical means should be taken to keep the oil from the belt. Where this is impossible the belt should always be removed from time to time and all oil extracted by some solvent such as naptha or benzine. Packing the belt in dry sawdust, whiting, or some similar absorbent material will sometimes answer the purpose. If it is impossible to remove the belt, wiping it while on the pulley with a dry cloth or waste will help. An excess amount of oil on a belt gives a bad frictional surface and causes it to slip, and also has a tendency to injure the sticking qualities of the ordinary cements in belt laps.

Should laps begin to open up on account of the presence of oil, it will be necessary to de-grease the parts to be joined and scrape off all old cement, or the new cement will not stick satisfactory. Do not think that you can remedy the trouble by riveting or driving tacks or belt fasteners through the joint. This simply makes a bad situation worse, for the leather will probably break where the metal pierces it. There is a right and a wrong way to repair laps, as well as to cement new laps, and the quickest and easiest way is not always the most economical in the end. It is easy to drive rivets or fasteners in a belt but it is just as well, before doing so, to think how much you weaken the belt at that point, how out of balance you make it, and consequently how it will jump every time the fasteners go around the cutterhead pulley. The proper way is to clean the open laps and apply good cabinetmaker's glue or regular belt cement, rub out all surplus glue, and then put on the clamps or nail a piece of board over the joint and let it stand at least an hour before releasing the pressure.

To make a new lap joint in a single-ply leather belt, square the ends and lay off the length of lap equal to about the width of the belt. Make the lap joint point the same way as other lap joints in the belt. Work each lap to a feather edge with a belt or spoke shave and scraper. Both

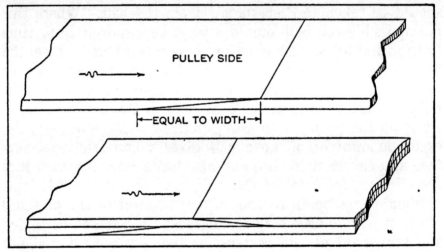

Fig. 85. Showing how the laps in single and double-ply belts should run.

laps must be made square and even so the joint will be the same thickness as the rest of the belt. Rough the surface of each lap with a rasp or piece of coarse sandpaper to give the cement a better chance to stick. Clamp or nail the belt to a smooth board so that it will lay perfectly straight when the laps match one above the other. If the work is being done in a cold room, warm the board and belt laps before applying glue or cement. When belt cement is used, follow the directions accompanying it, but in using cabinetmaker's glue, brush it on both laps and rub it in, then fold laps together and rub the outside of the joint briskly with some blunt tool to drive out all air bubbles and pockets. Then apply a flat even pressure by clamping or nailing a piece of board directly over the joint. Let stand over night, or half a day if possible, before releasing the pressure and putting the belt in service.

Avoid putting belts on too tight. If a belt is put on and run too tight, it becomes overstrained and injured, excessive friction is produced in the bearings and there is danger of damaging the machine boxes. If the demands on a belt are unusually severe, owing to a regular line of heavy work, a "floating" tightener can be placed on the slack side near the drive pulley. A yielding tightener, deriving its tension from its own weight or from coil springs, is a good thing, but when an unyielding tightener is used there is always a temptation on the part of the operator to screw it down too far and thereby make the belt altogether too tight. Lubricating oil, water, fine sawdust, and all general foreign substances should be kept off of cutterhead belts. Keep the belts clean but do not allow them to become dry and hard from want of proper lubrication. Leather belts in particular must be kept pliable and soft. An occasional but very light application of warm tallow, neatsfoot oil, or good belt dressing is just the thing to lubricate leather belts. The practice of doping a belt with rosin, soap, varnish, etc., to make it pull better is only a temporary relief and if continued to any extent will certainly ruin the best belt ever made. It is like doping the human body or mind to produce greater activity. The effects of a treatment soon wear off, leaving one in worse condition than before. If the stimulation is continued a collapse is inevitable.

It is a good plan in factories and mills where fast-feed machines are kept in continuous service, to have duplicate machine belts cut to length and ready for instant replacement. Belts should be inspected regularly and given proper care and attention, rather than be allowed to run on and on until they get in such a condition that some extreme action is necessary to save them from ruin.

Never throw a cutterhead belt off or on the pulleys at full speed; roll it on by hand or throw it on at very slow speed. When a belt is thrown onto a pulley moving at high speed, one edge of the belt is given a terrific strain all in an instant, and as a result that edge is stretched more than than the other. Often the belt becomes permanently set in this unequally-stretched condition and runs crooked ever after. Instead of traveling straight, the crooked belt occilates across the faces of the pulleys and is likely to set up end play in the cutterhead spindles.

A belt should be about 1-in. narrower than the pulleys over which it travels. If it does not seem wide enough to deliver the power required, put on a wider belt and use wider pulleys. A means sometimes employed to prevent slippage is to cover the pulley faces with leather. It is claimed that leather-covered pulleys will enable belting to transmit 25 per cent. more power than pulleys having a smooth iron surface. In preparing a pulley for a leather covering, begin by cleaning the surface thoroughly with naptha or benzine, then wipe it dry and warm it slightly. If possible, make the covering endless and about ⅛-in. to the foot shorter than the circumferance of the pulley. Place the endless cover on the pulley, pushing it on about 1-in. or more, then brush cement on the exposed inside surface of the cover and the exposed outside surface of the pulley, being sure that the cement is of the right consistency. Rub it thoroly into the leather and onto the pulley. Then take the pulley by its spokes and drive the cover on by striking it on the floor or bench. Do this quickly, but carefully, for if you strike too hard, or on one side more than the other, you may bend the leather so that it will be impossible to drive it onto the pulley. If it sticks a little, pry it loose with a screw-driver, then force it down, using the screwdriver as a lever. When the cover is completely on, rub the edges with some blunt tool to make good contact and work out surplus cement and any air pockets. A few rivets are generally added as a matter of safety.

To find the surface speed of a belt, multiply the diameter of a pulley over which it runs by 3.1416 or 3 1/7, and this product by the r.p.m. of said pulley. The surface speed of a belt should never exceed 5,000-ft. per minute. The speed of pulleys or any rotating parts can be easily and quickly figured if one remembers this simple fact: That the speed of the driving pulley in r.p.m. times its diameter equals the speed of the driven pulley in r.p.m. times its (the driven pulley) diameter. Example:

Speed of lineshaft............300 r.p.m.
Diameter of lineshaft pulley... 30-in.
Diameter of driven pulley.... 10-in.
Speed of driven pulley........   N

$$300 \times 30 = 10 \times N$$
$$9000 = 10 \times N$$
$$\frac{9000}{10} = N \text{ or } 900$$
$$300 \times 30 = 10 \times 900$$

Note: N represents unknown quantity.

A single-ply leather belt, 1-in. wide, running 800-ft. per minute will transmit 1 h. p. and for other thicknesses figure the unit speed as 500, 400, and 300, respectively, for two-ply, three-ply and four-ply. Remember, the greater the speed of the belt, the more horse power transmitted in direct proportion, but do not let the speed exceed a mile a minute.

To figure the length of belt for a drive when it is inconvenient to measure the distance with a tape line: Add the diameter of the two pulleys, multiply this result by 3.1416, and divide by 2. To this quotient, add twice the distance between centers of shafts and this will give the required length provided both pulleys are about the same size. When one pulley is considerably larger than the

other, square the distance between the centers of the shafts;
add to this the square of the difference between the radii of
the two pulleys; from this total extract the square root and
multiply by two. Call the total thus obtained, T. Then
add the diameters of the two pulleys together, multiply
result by 3.1416 and to one-half of this add the amount
just designated as T, and you will have the length of belt
required. To find the number of lineal feet in a roll of
belting of any kind or ply, add the diameter of the roll
in inches to the diameter of the center hole, multiply by
the number of coils, counting from the center to the outside, and multiply this product by .1309.

SETTING A MOLDER.

A molder, like an engine or other important piece of
machinery, should set level and be firmly bolted down to a
solid floor or foundation. Concrete foundations are good
and unless there is a good concrete or heavy plank and
timber floor, a special foundation should be prepared. The
conventional method of bolting a machine to a plank floor
or timber foundation is to use lag screws and washers.
Lag screws are also used to anchor machine bases on concrete floors and foundations. Holes are cut in the concrete
to correspond, in position, to holes in the machine base.
The machine is then positioned over the holes, lined up
with the driving shaft, if belt drive is to be used, and
leveled with hardwood shims so all feet rest firmly on the
foundation. Small channels are then cut to each hole in
the concrete and when the lag screws are hung in place
in the machine-base holes, melted babbitt or lead is poured
in around them. This anchors the machine very substan-
tially.

Common sulphur is sometimes used in the place of babbitt metal because of its comparative cheapness. When sulphur is used, no channels need be cut for pouring because
the melted sulphur can be poured directly into the holes

and the lag screws set in immediately after. This is possible because sulphur cools more slowly than metal. Of course, the lag screw is not pushed all the way down in the cooling sulphur, but within about ¾-in. or 1-in. of its limit, and turned the balance of the distance with a wrench. When the sulphur hardens it holds the lag screws firmly in place.

Another method of fastening machines to concrete foundations is to put down inverted anchor bolts with large washers (washers which prevent bolts from turning) and fill in around these bolts with cement, sulphur, or melted metal. The machine base is then set down over the projecting ends of these bolts and fastened down with nuts and washers.

If a molder is to be driven by an individual motor, the usual practice is to connect the motor directly to the machine countershaft by a flexible coupling. A flexible coupling is preferred to a solid coupling for several reasons; it relieves the motor of certain mechanical jarring and aids it in getting under way to better advantage when started; also permits of a slight misalignment in the shafts without causing the usual trouble incident to using solid couplings.

The size motor required to drive any particular size molder depends largely upon the kind of service the machine is to be put to, whether light, medium, or heavy use, intermittent or continuous, slow-feed or fast-feed service, etc. The following table shows pretty nearly the size motors required for various molders operated under average conditions.

Size of the Molder	Light and Medium Service	Heavy Service, Fast Feeds, or Both
6-in.	5 h.p. motor	7½ to 10 h.p. motor
8-in.	7½ h.p. motor	10 to 15 h.p. motor
10-in.	10 h.p. motor	15 to 20 h.p. motor
12-in.	15 h.p. motor	20 to 30 h.p. motor
15-in.	20 h.p. motor	25 to 30 h.p. motor
18-in.	20 h.p. motor	25 to 35 h.p. motor

CHAPTER XVI

MOLDING SHAPES.

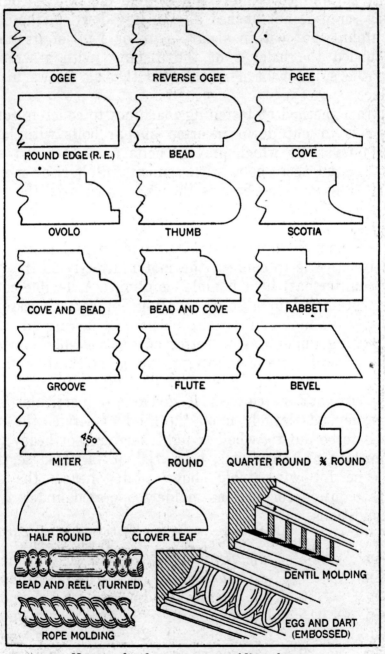

Names of a few common molding shapes.

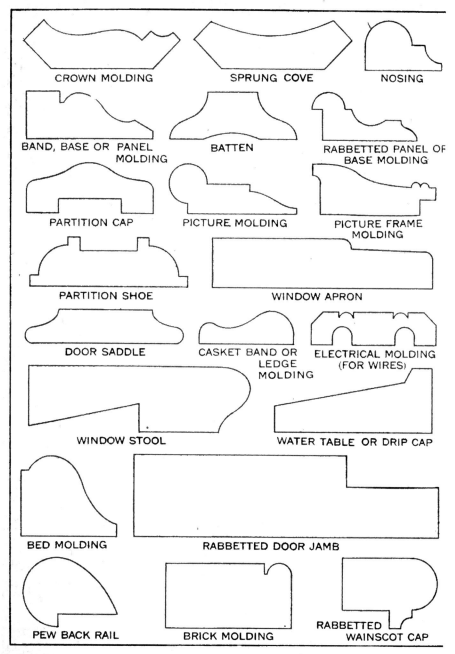

Showing one style each of several different kinds of molding commonly used in the building trades and casket manufacturing.

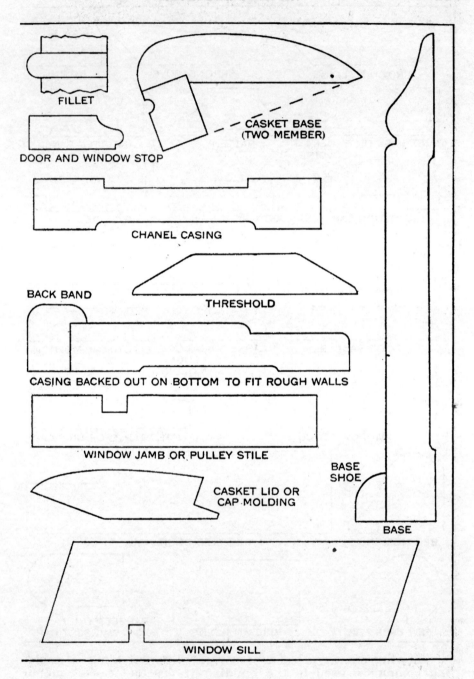

FILLET

CASKET BASE
(TWO MEMBER)

DOOR AND WINDOW STOP

CHANEL CASING

THRESHOLD

BACK BAND

CASING BACKED OUT ON BOTTOM TO FIT ROUGH WALLS

WINDOW JAMB OR PULLEY STILE

CASKET LID OR
CAP MOLDING

BASE
SHOE

BASE

WINDOW SILL

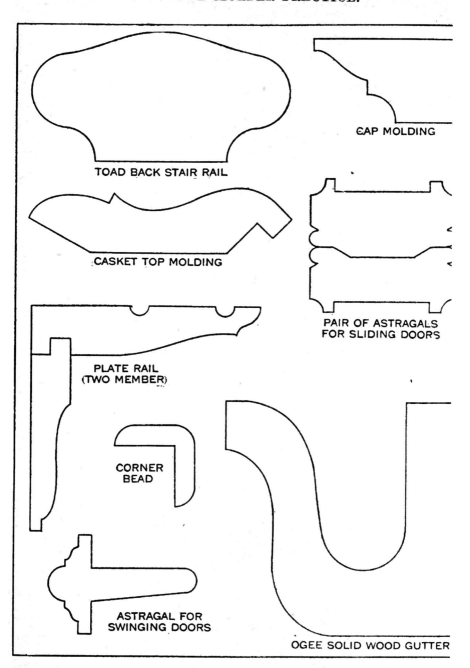

TOAD BACK STAIR RAIL

CAP MOLDING

CASKET TOP MOLDING

PAIR OF ASTRAGALS
FOR SLIDING DOORS

PLATE RAIL
(TWO MEMBER)

CORNER
BEAD

ASTRAGAL FOR
SWINGING DOORS

OGEE SOLID WOOD GUTTER

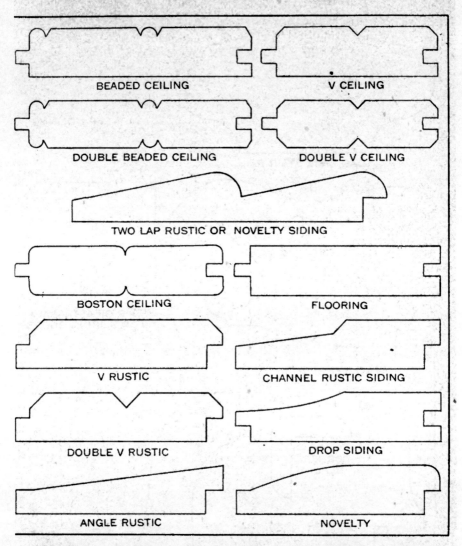

BEADED CEILING

V CEILING

DOUBLE BEADED CEILING

DOUBLE V CEILING

TWO LAP RUSTIC OR NOVELTY SIDING

BOSTON CEILING

FLOORING

V RUSTIC

CHANNEL RUSTIC SIDING

DOUBLE V RUSTIC

DROP SIDING

ANGLE RUSTIC

NOVELTY

The above patterns are generally run on fast-feed molders or matchers, the V's, beads and bevels being worked with profile discs mounted on the profile spindle. They can also be made with ordinary knives on four-side, square-head molders.

A general purpose molder which can be fitted with any kind of slip-on head and used for either odd, detail or fast-feed work.

A completely motor-driven outside molder: size six inch. Each cutterhead and the feed works individually driven thru five independently controlled motors. Cutterhead spindles run in ball bearings and take slip-on heads for either slow or fast-feed work. Invented by R. D. Eaglesfield, Indianapolis, Ind.

INDEX TO ILLUSTRATIONS.

CHAPTER VI.

CHAPTER VII.

CHAPTER VIII.

CHAPTER XIII.

REPRESENTATIVE TYPES
OF MODERN
MOLDING MACHINES

□ □
□

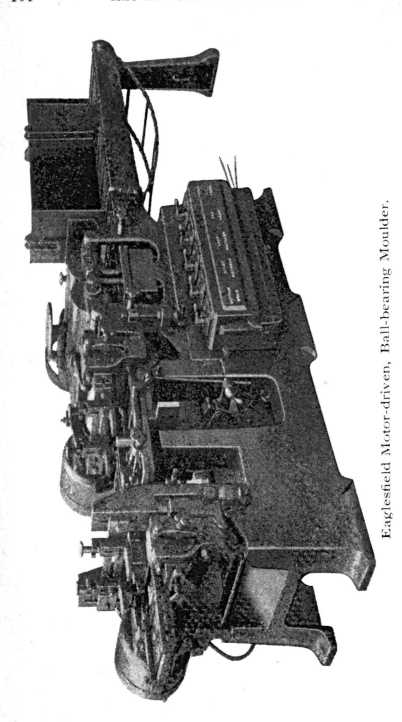

Eaglesfield Motor-driven, Ball-bearing Moulder.

The PATENTED
EAGLESFIELD MOULDER
COMPLETELY MOTOR DRIVEN

SPECIFICATIONS

SIX-INCH MOULDER.

Diameter of Spindles, where heads slip on.............1-13/16-in.
Diameter of Cutter Heads ....4 to 7-in.
Type of Cutter Heads....................Slip-on, Round or Square
Spindle Speeds ....3,450 to 3,600-r.p.m.
Vertical Spindles (length for head).......................4-in.
Vertical Spindles, maximum angle ....45-deg.
Number of Feed Speeds 4
Standard Feeds ....25 to 100-ft. per min.
Diameter of Feed Rolls.......................................8-in.
Width of Stock ....00 to 6-in.
Thickness of Stock ....00 to 4-in.
Current.............110, 220, 440, 550 Volt, 2 or 3 Phase, 60 Cycles
Table Height ....34-in.
Length over all ....88-in.
Width over all ....46-in.
Approximate Weight ....6,800-lbs.

TWELVE-INCH MOULDER.

Diameter of Spindles, where heads slip on.............1-13/16-in.
Diameter of Cutter Heads...........................4 to 7-in.
Type of Cutter Heads....................Slip-on, Round or Square
Spindle Speeds ....3,450 to 3,600-r.p.m.
Vertical Spindles (length for head).......................4-in.
Vertical Spindles, maximum angle ....45-deg.
Number of Feed Speeds 4
Standard Feeds ....25 to 100-ft. per min.
Diameter of Feed Rolls ....8-in.
Width of Stock ....00 to 12-in
Thickness of Stock ....00 to 4-in.
Current..............110, 220, 440, 550 Volt, 2 or 3 Phase, 60 Cycles
Table Height ....34-in.
Length over all ....96-in.
Width over all ....48-in.
Approximate Weight ....8,000-lbs.

EXTRA ATTACHMENTS.

Hopper Feed Attachment to Take.................3 or 6-ft. Lengths
Jointers for Top, Bottom and Side Heads.
Knife Grinder, motor driven.
Shaving Hoods.

VONNEGUT MACHINERY COMPANY
INDIANAPOLIS, U. S. A.

Mattison Motor-driven. Ball-bearing. Heavy-duty Moulder.

Hermance No. 50 Molder—8, 10 and 12-inch

HERMANCE MOLDERS

Equipped either for general
purpose or fast-feed work.

Hermance No. 40 — 12 and 16-inch.

A heavy molder that combines the desirable
features of both the inside and outside types. No
belts on working side allows free access for quick
adjustments and easy set-ups. Has the massive
frame, rigid bed and heavy bearings on both sides of
top and bottom cutterheads and feed rolls that are
the desirable features of the true inside molder.
Hopper feed if desired.

Hermance No. 50 — 8, 10 and 12-inch.

Noted for remarkable ease and speed with which
set-ups may be made, due to its accessibility and
the simplicity and completeness of adjustment
features, making it an exceptional machine for
general purpose work. May be equipped with
our fast-feed features, which are of unusual simplicity.

Hermance 6-inch — one, two, three or four sided.

An ideal machine for accurate working of small
moldings.

Write for catalog illustrating and fully describing our line of modern molders and other
high-grade wood-working machinery.

HERMANCE MACHINE CO.
Williamsport, Pa.

Woods No. 419 Super-six Moulder.

Woods "Sawco" Electric Hand Grinder.

The Woods "Sawco" Electric Hand Grinder in Operation.

WOODS "SAWCO" ELECTRIC HAND GRINDER

ONE OF WOODS LABOR AND TIME-SAVING DEVICES

Adapted Especially for Moulding Machine Knives

On fast-feed work, where formed knives are used, this grinder is indispensable. It will eliminate almost 50% of the time lost in taking care of the cutters. After a formed cutter has been jointed several times, it becomes necessary to remove the cutter-head from the machine and take the cutters out and grind off the heel. This not only takes a great deal of time in itself, but means an equal loss of time in re-setting the knives in the head and getting the head back into the machine. In addition to this, more jointing is necessary before the head can be put in operation with the consequent loss of time and expensive high-speed steel. The SAWCO grinder will eliminate all of this and the heel can be ground off of the knives without taking the knives out of the head or even removing the head from the machine or disturbing the set-up in any way.

S. A. WOODS MACHINE COMPANY
BOSTON, U. S. A.

The History of the Moulder is the History of the S. A. Woods Machine Co.

ImTheStory.com

Personalized Classic Books in many genre's

Unique gift for kids, partners, friends, colleagues

Customize:

- Character Names
- Upload your own front/back cover images (optional)
- Inscribe a personal message/dedication on the
  inside page (optional)

Customize many titles Including
- Alice in Wonderland
- Romeo and Juliet
- The Wizard of Oz
- A Christmas Carol
- Dracula
- Dr. Jekyll & Mr. Hyde
- And more...

Lightning Source UK Ltd.
Milton Keynes UK
UKOW051828271212

204136UK00009B/581/P